Modern Physics Experiments
for the
High School

Modern Physics Experiments for the High School

William K. Esler

Parker Publishing Company, Inc.
West Nyack, N.Y.

© 1970 BY
PARKER PUBLISHING COMPANY, INC.
WEST NYACK, N.Y.

ALL RIGHTS RESERVED. NO PART OF THIS
BOOK MAY BE REPRODUCED IN ANY FORM, OR
BY ANY MEANS, WITHOUT PERMISSION IN
WRITING FROM THE PUBLISHER.

LIBRARY OF CONGRESS
CATALOG CARD NUMBER: 79-117211

PRINTED IN THE UNITED STATES OF AMERICA
B&P—0-13-597005-9

Dedication

*To Mary,
whose dedication to this task
sometimes exceeded that of my own.*

How This Book Can Give New Meaning to Your Physics Laboratory

This book is directed to the many physics teachers who have the uncomfortable feeling that the quality of the laboratory portion of their course may not match the high standards of their classrooms. Hopefully, information presented here will help those teachers to reorganize their laboratory methods and reorient their thinking so that this very important part of students' training in science will take on new vitality and purpose.

Physics instruction in the high school has become associated, in the minds of many, with the vocational goals of a relatively small group of students. These students are thought to possess above-average academic ability and interest in science. A criticism of some of the new curricula is that, through attempting to emphasize the structure of science, they have become so theoretical that they indeed tend to attract only this student stereotype. This situation only enhances the erroneous image from which physics instruction now suffers. Many physics teachers and professional physicists feel that the study of their discipline has much to offer to the liberal education of all students, regardless of future vocational plans. The development of a questioning and rational spirit, the fulfillment of aesthetic and emotional needs through the appreciation of creations, and the acquisition of a faith that order exists in our universe are examples of the liberal values which may be enhanced through science training. The task set before physics teachers appears to be to build a physics course that emphasizes such values while retaining something of the structure of the discipline, and at the same time not neglecting entirely the technology that renders the science useful to mankind. In this way a greater number of students with a broad spectrum of abilities and interests may be attracted to enroll in this science.

It need not be necessary to throw out existing programs to realize some of these goals. In the laboratory it becomes not so much a question of which exercises are incorporated in a program of study as it is the goals which decide the nature of their presentation. Attempts at developing trouble-free laboratory programs—programs teachers find easy to administer—have often resulted in detailed, step-by-step instructions for the students that have relieved them of the responsibility for making even the smallest decisions. The students, of course, react to this lack of involvement in wholly predictable ways. They simply go through the motions of following directions with little understanding of the basic principles involved. Thus, they develop distorted views of the attitudes and skills required to deal with science. The same basic laboratory exercises that result in these sorts of learning outcomes may be reorganized so as to cause realization of objectives much more positive in nature. With proper format, traditional laboratory experiments can be made the vehicles to carry students to a real understanding of the processes of science—and these, properly structured, become the means of developing in the students the true skills of the scientist.

Attempts at defining the skills of the scientist have led the Commission on Science Education of the American Association for the Advancement of Science to adopt the following list of processes:

Primary Processes—1) Observing 2) Classifying 3) Measuring 4) Communicating 5) Inferring 6) Predicting 7) Recognizing Time/Space Relations 8) Recognizing Number Relations.

Integrated Processes—1) Formulating Hypotheses 2) Making Operational Definitions 3) Controlling and Manipulating Variables 4) Experimenting 5) Interpreting Data 6) Formulating Models. These skills are believed to be accumulative, at least to the degree that the integrated processes build upon prior facility with the primary processes.

It is the purpose of all activities described in this book to introduce and reinforce one or more of the process skills. After engaging in a number of laboratory exercises of this design, the learner will have acquired a basis for understanding the real tactics of dealing with problem-solving situations. He will think of the physics laboratory as more than a place where he tinkers with some weights and pulleys, and fills in empty spaces in a laboratory manual. He will feel the confidence that comes with increased skill in dealing with problems, and he will know the thrill of devising experiments to test his own hypotheses. He will have experience in formulating hypotheses by induction and extrapolating others by deduction. Just as importantly, his teacher will know the increased satisfaction that comes from dealing with his students in a new and different role—that of consultant and partner in the learning process.

There is a challenge for the teacher in organizing his physics laboratory in the way suggested here. The challenge lies primarily in the increased teacher involvement necessary to putting such a program over. When students are allowed greater freedom in the pursuit of knowledge, it must be expected that they will require a greater amount of guidance from the teacher until they develop the proper confidence, attitudes, and skills necessary for independent operation. Only teachers willing to devote the time and energy required of this kind of teaching will ever know the satisfaction of seeing their students eventually operating as independent investigators who explore scientific principles while utilizing the intellectual tools with which he has endeavored to equip them.

ORGANIZING FOR INSTRUCTION

A good high school physics laboratory program may be operated with a number of formats. The suggestions which are offered in these remarks describe one laboratory program which has operated successfully. The program may be equally successful with many modifications of procedure and manipulations of the ordering of the exercises. The one ingredient that would appear to be required in its operation is the teachers' agreement with the basic goals of the high school physics laboratory operation. These are: 1) The students should acquire the basic knowledge and manipulative skills necessary to do experimentation in the laboratory, 2) the students should acquire the process skills that allow them to become more than merely manipulators of apparatus, and 3) students should acquire the ability and confidence to operate as independent investigators.

The organization of a high school physics program that attempts to meet the above stated goals is not difficult. Suggestions for that organization are: 1) arrange the laboratory exercises in blocks by topic or theme, 2) conclude each block with an exercise which requires the student to organize an independent investigation, and 3) evaluate the performance of each investigator, or team of investigators, by procedures designed to emphasize the process of laboratory investigation.

Arranging Laboratory Exercises by Blocks

Arrangements of science laboratory programs in our high schools are more often dictated by office scheduling practices and tradition than by sound content organization and learning theory. In many instances, laboratory periods are scheduled for specific days of the week regardless of the degree of relevance of the experiences to those taking place in the classroom.

An alternative to these practices is to schedule laboratory exercises in blocks. This means having laboratory each day, for five, six or ten days running. For the duration of the block, the students are assigned a number of laboratory problems, and when one problem is completed they proceed to the next. Since the students' work proceeds at different rates, the teacher must decide which exercises comprise the minimum block he would require.

The exercises that comprise the block are fairly well structured. They are designed to acquaint the students with laboratory apparatus and procedures, as well as to illustrate basic principles of physics. After completing the prescribed block of exercises, which is designed to strengthen such skills as measuring, manipulating variables, and interpreting data, the student is introduced to the challenge of an open-ended problem, or individual research project.

The argument that single periods of 40 to 45 minutes are too short to accomplish anything in a laboratory is valid only for very few exercises. It has been found that when the student is required to schedule his own activity, he generally makes the necessary adjustments in order to get the maximum production from any time allotment.

Limited Independent Investigation

Regardless of its organization, no program of laboratory exercises which directs the student toward known and expected outcomes is entirely effective in the development of process skills. It is only when the student feels the responsibility of clarifying a problem and choosing the procedures and apparatus required in its study that the learning situation achieves maximum effectiveness.

The most serious challenge to the teacher who hopes to guide his students toward acquiring something of the feel of scientific investigation is the development of the proper laboratory exercises. Some requisites for such exercises are: 1) The problem under consideration may be investigated in a reasonable number of class periods (perhaps three to five days); 2) the problem may be investigated with apparatus with which the learner is already familiar or may readily become familiar; 3) provisions are made for differences in the individual student's ability to cope with independent study; 4) the nature of the problem is such that most of the students may reasonably be expected to achieve a measure of success in its investigation.

Most of the requirements for limited independent investigations are met with the proper shaping of common laboratory exercises which illustrate principles not stressed in the classroom. Since no course of study may encompass all facets of a topic, such exercises are not difficult to find in high school physics manuals. These exercises, which have been geared to the high school student, ensure the relative simplicity of the activities and increase the likelihood that the problem may be investigated with the use of apparatus common to physics laboratories at that level. For maximum ease of administration, the problem to be investigated must be stated clearly by the teacher and related to concepts previously learned. The amount of guidance the teacher provides may be adjusted to the individual investigator. As a research director, the teacher should attempt to help the students progress in their investigations, but progress should not be achieved at the cost of denying the student the opportunity of structuring his own experiments and interpreting his own data. Procedural suggestions should always be made so that the students must choose between several alternatives. They must at all times feel their investigations are their own private realm, that both success and failure in the endeavor are attributable to their own actions.

Evaluation of Laboratory Performance

High school students quickly learn the things that their teachers consider important. Their judgment is normally based upon those things teachers use as criteria for issuing grades. What good is it for teachers to give constant lip service to student attitudes and skills in laboratory situations and yet evaluate performance on the basis of the ability to fill in blank spaces of a prepared worksheet; or worse yet, the practice of issuing grades on the basis of the number of questions of doubtful relevance that have been answered at the end of each written exercise? To emphasize the development of process and laboratory skills, the teacher must evaluate student performance using these same skills as performance criteria.

The proper evaluation of laboratory performance forces teachers into a situation which many purposely avoid—the making of subjective judgments of student performance. No written test has yet been devised that measures the development of the abilities to formulate hypotheses, manipulate variables, and others of the process skills. It is doubtful that a written test may ever be devised to do this job for the teacher, for the advantage on any written test goes to the student who has performed a preponderance of written work in the course of his daily activities. Though test items exist that call for such skills as interpretation of artificial data and the formulation of decisions as to the validity of inferences, such items possess only limited reliability in the context of a real problem-solving experience. The only valid means of assessing a student's ability to define problems, formulate hypotheses, and carry on the research required to test his hypotheses is to observe his behavior in a practical situation. For this reason physics teachers should provide the students with the opportunity to funcion in at least part of their laboratory work in relatively unstructured situations; and these same teachers must be willing to accept the responsibility for subjective evaluation of those student activities that are truly meaningful in terms of his teaching goals.

One of the major advantages of unstructured laboratory activity is that it provides increased opportunity for one-to-one conversations between teacher and pupil. This author feels it would be a definite deterrent, to establishing the best possible relationships with his students, for a teacher to attempt to overtly evaluate student performance during consultation periods. It is through this sort of relationship, however, that the teacher comes to know the real abilities and level of performance of the individual students. Judgments arising from this kind of intimate relationship provide the teacher with his best criteria for determining the success of his program.

To ensure that all students have been engaged in a discussion during a single laboratory unit, the teacher may wish to use the last two or three laboratory periods for scheduled discussion sessions. The time factor usually necessitates that he talk with three or four students at a time, thereby reducing to some extent the level of personal interaction. One benefit of small group discussion is that the students may evaluate the quality of their own responses in relation to those of the others in the group.

The sort of question that should be asked of the students follows the general pattern: What is the problem you are investigating? Why did you proceed in this way? Can you explain your interpretation of data?

Notice that the burden placed on the student is to clarify, defend, and explain his experimental design. Any statement of "wrongness" of his data should be avoided, for the fact remains that his data are never wrong. In the event it differs from expected values it is the right product of a faulty procedure. By emphasizing procedures and skills the teacher makes the student aware that these sort of things are of importance. By deemphasizing "right" answers, the teacher automatically impresses upon the student that physics is a product of investigation—a dynamic process, not a stagnant pool of established fact. The guiding dictum must be: I will evaluate the process of scientific investigation first and its product second.

A procedure the physics teacher may find useful for the evaluation of the laboratory block is the *contract*. The contract is an organizational method wherein the student agrees to satisfactorily complete a specified number of exercises to earn a particular letter grade. A contract may be entered into in a formal way; that is, the student may sign an agreement that prescribes the grade for which he is striving and the work that is required of him for its issuance. Less formally, the teacher may simply specify for a class the requirements for achieving each letter grade. In nearly all instances of use of the contract method, student performance is judged only as being satisfactory or unsatisfactory. The contract permits the student to some degree to select the grade for which he will strive. However, it is often necessary for the teacher, toward the end of a block of experiments, to inspire some of his charges to seek the highest possible mark. The contract is no cure-all for motivation problems. Its use, however, eases the task of laboratory evaluation and strengthens the teacher's position of fostering an atmosphere of maximum independence for students. A concomitant benefit of the contract system is that the students appear to become aware that the development of a certain number of laboratory skills is of some importance. This concept is then further reinforced with the assignment of the open-ended problem at the completion of the laboratory block.

THE FORMAT OF THE TEXTUAL MATERIAL

The exercises in *Modern Physics Experiments for the High School* are arranged in units or blocks. Following each block of exercises are found several suggestions for problems suitable for independent investigation. No single laboratory exercise need be considered a requirement for students to successfully deal with the problem-solving exercise. Each exercise may be assigned according to the judgment of the teacher.

The exercises begin with a statement of the rationale for using that particular activity. In this rationale may be found a description of the process skills thought to be reinforced therein. Following the statement of rationale and headed by the notation "(Student)" is an introductory paragraph and description of procedures that students may follow in completing the exercise. These remarks and procedures are directed to the students and are written in a language appropriate to their direct application in the high school physics laboratory. Optional activities appear at the end of many of the exercises. Such options may be used to allow for individual differences in the rate at which students work or to strengthen the skills of all students. Where possible, answers to questions posed in the exercise will appear at the end of the exercise. Where answers are not feasible, the reader will find explanations of the various results students may be expected to obtain in carrying out the procedures. Statements of caution to aid students in avoiding some common errors in laboratory technique are found throughout the procedural directions.

W.K.E.

Acknowledgments

The author wishes to express his appreciation to the physics students of Ellet High School, Akron, Ohio, who suffered through the early attempts to restructure their physics laboratory and provided invaluable commentary concerning the understandability of the exercises.

Also a great deal is owed by the author to Dr. S. J. Mark of Kent State University, who helped the author to understand the mechanics of developing a manuscript and expressed confidence in the outcome.

The illustrations appearing in this book are the work of Miss Mary Reed of Orlando, Florida.

Table of Contents

Unit I   Measurement and Forces   21

Exercise 1—The Metric System • 23
Exercise 2—The Vernier and Micrometer Caliper • 27
Exercise 3—Density of Solids • 31
Exercise 4—Vector Forces • 33
Exercise 5—Resolution of Forces • 37
Exercise 6—The Law of Moments • 43
Exercise 7—A Non-Uniform Lever • 46
Exercise 8—Parallel Forces • 49
Exercise 9—Optional Problems to Test Laboratory Skills • 52

Unit II   Mechanics of Solids   57

Exercise 10—The Pendulum • 59
Exercise 11—Pulley Systems • 62
Exercise 12—Friction • 70
Exercise 13—The Inclined Plane • 73
Exercise 14—Hooke's Law • 77
Exercise 15—The Wheel and Axle • 80
Exercise 16—Velocity and Acceleration • 83
Exercise 17—Conservation of Momentum • 86
Exercise 18—Optional Problems to Test Laboratory Skills • 88

Unit III   Mechanics of Liquids and Gases   93

Exercise 19—Pressure of Liquids • 95
Exercise 20—The Law of Buoyancy • 97
Exercise 21—Density and Specific Gravity • 100
Exercise 22—Specific Gravity of Liquids • 105
Exercise 23—Atmospheric Pressure • 108
Exercise 24—Optional Problems to Test Laboratory Skills • 112

Unit IV   Wave Phenomena—Sound and Light   117

Exercise 25—Frequency of a Tuning Fork • 119
Exercise 26—Determination of the Wavelength and
               Velocity of Sound by Resonance • 122
Exercise 27—Vibrations of Strings • 125
Exercise 28—Intensity of a Light Source • 129
Exercise 29—Reflections from a Plane Surface • 132
Exercise 30—Images in a Curved Mirror • 137
Exercise 31—Refraction of Light • 141
Exercise 32—Refraction in a Convex Lens • 146
Exercise 33—The Visible Spectrum • 149
Exercise 34—Waves in a Ripple Tank • 152
Exercise 35—Optional Problems to Test Laboratory Skills • 156

Unit V   Static Electricity and Direct Current   161

The Use of Meters • 163
Exercise 36—Static Electricity • 165
Exercise 37—Ohm's Law and Resistance • 168
Exercise 38—Cells in Series and Parallel • 171
Exercise 39—The Electrolytic Cell and
               the Activity Series • 175

Contents

Unit V Static Electricity and Direct Current *(Continued)*

Exercise 40—The Lead Storage Cell • 178
Exercise 41—Electroplating • 182
Exercise 42—Optional Problems to Test Laboratory Skills • 185

Unit VI   Circuits and Magnetism   *191*

Exercise 43—The Wheatstone Bridge • 193
Exercise 44—Series Circuits • 196
Exercise 45—Parallel Circuits • 200
Exercise 46—Series and Parallel Circuits Combined • 203
Exercise 47—Electrical Power • 206
Exercise 48—Magnetic Fields • 208
Exercise 49—Magnetic Fields About
            Current Bearing Conductors • 211
Exercise 50—Induced Currents • 214
Exercise 51—The Transformer • 216
Exercise 52—Inductance and Capacitance • 220

Unit I

Measurement and Forces

The Metric System
The Vernier and Micrometer Caliper
Density of Solids
Vector Forces
Resolution of Forces
The Law of Moments
A Non-Uniform Lever
Parallel Forces
Optional Problems
    to Test Laboratory Skills

Unit I

Measurement and Forces is designed as an introductory unit. The activities required of the student are designed to be relatively simple yet will lead them to the beginnings of the proper attitudes and skills required for the successful handling of independent investigation.

EXERCISE 1—THE METRIC SYSTEM

APPARATUS

1. A meter stick calibrated with millimeters
2. An English Ruler
3. A graduated cylinder
4. Platform or triple beam balance
5. A set of metric weights
6. Eye dropper
7. Quart Measure

(Teacher) This exercise on the metric system serves a fourfold purpose. It serves to acquaint the student with this most important system of weights and measures, and at the same time causes him to become aware of the limitations of the accuracy of measuring devices. He learns to provide acceptable data by the time-honored system of averaging a number of reasonable results, and to manipulate common measuring devices. The Optional Exercise involves estimating lengths and weights in metric units and serves to provide the student with points of reference in his efforts to accommodate the metric system to his knowledge of the English system.

(Student) The metric system is universally used in scientific laboratories. This widespread use requires of all students that they become very familiar with it. The common metric units for length are the millimeter (mm), centimeter (cm), meter (m), and kilometer (km). For weight, two mass units, the gram (g) and kilogram (kg) are commonly employed, and for volume the milliliter (ml), cubic centimeter (cc), and liter (l). Study the table of metric measures in your textbook and establish in your own mind its relationship to the English system. (Teacher: Be sure the students understand that though metric weight is often expressed in miss units the proper force units of the metric system are dynes and newtons.)

$$\text{Force (weight) in newtons} = \text{mass in kilograms} \times \left(\frac{9.80 \text{ newtons}}{\text{kilogram}}\right)$$

$$\text{Force (weight) in dynes} = \text{mass in grams} \times \left(\frac{980 \text{ dynes}}{\text{gram}}\right)$$

PROCEDURE

Part A—Measures of Length

Examine a meter stick. The smallest division on the meter stick is the millimeter. Ten millimeters are contained in one centimeter. There are one hundred centimeters in the length of the meter stick. What is the greatest accuracy possible to obtain with a meter stick?

When each member of the investigating team believes he understands the graduations on the meter stick he should measure the length of the laboratory table to the nearest millimeter and write his results on a piece of paper. *Do not let the other investigators see your results.* With at least three or four members in the comparison group show your results.

1. How do the results of your measurements compare?
2. Which investigator is most correct?

In scientific investigation when the results of independent measurements are not the same, an accepted value is obtained by averaging all reasonable measurement data. A single investigator will take the same measurement a number of times (three is normally considered a minimum) and average his results to obtain an accepted value. Apply this guideline to all measurements called for in this exercise and all subsequent exercises in the physics laboratory. Measure the length and width in both the English and

The Metric System

Metric systems of a table or desk top. (Teacher: At this point you may wish to refer the students to a discussion of significant figures in the textbook.)

Accepted Value

3. The length of the table = _____ cm or _____ in.
4. The width of the table = _____ cm or _____ in.
5. The area of the table = _____ cm² or _____ in.²

Now measure the length of your textbook or lab manual in both systems.

6. The length of the book is _____ cm or _____ in.
7. Can you use the data concerning the length of the textbook to determine the number of centimeters in one inch?

Part B—Measures of Mass and Weight

What is the difference between the mass and the weight of an object? Quite simply weight is a measure of the attractive force between the object and the earth. An object's weight will vary according to the location. Mass is a measure of the inertia of an object and this concept is a constant for any location on earth or in space. The variations of weight on the surface of the earth are small enough that they are ignored for most purposes and the units of metric mass are generally used to describe weight measurements. In using mass units as measures of weight it would be more correct for an observer to talk about the gravitational forces exerted upon the mass unit rather than just naming the mass unit. For reasons of efficiency, however, mass units are often named as weight units. The two concepts should be clearly differentiated and firmly fixed in the mind of the experimenter to avoid any confusion.

The smallest measures of metric weight normally used in the high school physics laboratory are tenths and hundredths of the gram. The kilogram is equal to 1000 grams.

To gain some appreciation of the magnitude of the gram as a unit of weight, tear a sheet of notebook paper into four equal parts. Adjust the knurled set screw on a platform or triple beam balance until the pointer remains at zero or swings an equal number of divisions on each side of the zero mark. Now set the counterweight on the balance at 1 gram.

1. How many quarters of the notebook paper must be placed on the balance to equal or exceed one gram of weight?
2. Now find an accepted value for the weight of a meter stick; a book; a shoe. (Did you make at least three separate measurements of each object?)
3. If there are 454 grams per English pound find the weight of these three objects in pounds.

Part C—Measures of Volume

The smallest unit of metric volume normally considered in a high school physics laboratory is the milliliter. There are 1000 milliliters in one liter.

Secure a graduated cylinder and fill it partially full of tap water. Hold the graduated cylinder at eye level and read the lowest point of the curved water surface. Now have two other investigators do the same.

1. Are the readings the same by all investigators?
2. How may you determine an accepted value for the volume of water in the graduated cylinder?

Now add water drop by drop with an eyedropper until one milliliter is added to the previous volume.

3. A milliliter is how many drops of water?

Now fill the graduated cylinder as many times as is necessary to put 1000 ml. of water in a quart container.

4. What is the approximate relationship between 1000 ml. of volume and a quart of volume?

Part D—Teacher's Option

Guess the length in metric units of some common objects such as a pencil, a finger, an arm, etc., and compare with the measured values. Do the same with metric units of weight using such items as a book, a ring, a quarter, etc.

DISCUSSION OF OUTCOMES

Part A

1. The results of student measurements will normally differ by a factor of several millimeters.

The Vernier and Micrometer Caliper 27

2. No one investigator's measurement may be considered right.
3. Any reasonable value derived by measurement.
4. Any reasonable value derived by measurement.
5. The product of the accepted values of lengths and widths from items 3 and 4.
6. Any reasonable value derived by measurement.
7. Approximately 2.54.

Part B

1. 2 or 3 quarters of a notebook page should tip the balance.
2. Any reasonable value.
3. The weight of each object in pounds is found by dividing its weight in grams by 454.

Part C

1. Volume readings by the investigators will normally vary plus or minus one ml.
2. An accepted value is obtained by averaging three or more separate volume readings.
3. It takes approximately 20 drops of water to equal one ml. of volume.
4. 1000 ml. (1 liter) of volume is equal to approximately 1.1 quarts.

EXERCISE 2
THE VERNIER AND MICROMETER CALIPER

APPARATUS
       1. Vernier Caliper
       2. Micrometer Caliper
       3. Hollow Cylinder

(Teacher) This exercise is designed to extend student skills in making measurements; to cause him to recognize the limitations of accuracy of single measurements; and to add two new measuring devices to his list of available physics laboratory tools.

(Student) The accuracy of taking measurements directly from a rule is necessarily limited to the smallest division readable. Two instruments have been devised, and are widely used, which increase the ability of an observer to accurately measure small lengths. The principles and use of the vernier and micrometer calipers are the subjects of this experiment.

PROCEDURE

Part A—The Vernier Caliper

Examine a vernier caliper closely. When the jaws of the caliper are closed the left or zero side of the sliding scale lines up with the zero on the fixed scale. If your caliper has both the English and Metric scales look only at the lower or Metric scale. When an object is placed between the jaws of the calipers the sliding scale moves over the fixed scale and the object's dimension may be read as to whole *millimeters* by reading on the fixed scale directly above the zero index of the sliding (or vernier) scale. To determine tenths of a mm. locate the one line of the sliding vernier scale which exactly coincides with a mm. line on the fixed scale. The number of this line on the vernier (0-10) is the number of tenths of a mm. to be added to the whole millimeters to thus record a measurement to the nearest .1 of a millimeter.

When each of three or four investigators feels he understands the procedures for using the vernier caliper let him find the thickness of some common metal object such as a nail, a metal rod or metal sheet and record the results on a sheet of paper. *Do not permit the other investigators to see the results.* Now compare results.

1. Are the measurements the same?
2. Which investigator is most correct?

Obtain an accepted value for the thickness of the object by finding the average of the results obtained by the separate investigators. Now each investigator should find an accepted value for the thickness of the worktable top.

3. How do you determine the accepted value?

At the top of the outside caliper jaws you will see a second set designed to measure inside dimensions of cylinders. The scale

The Vernier and Micrometer Caliper

is read the same as above. Find an accepted value for the inside diameter of a cylinder.

4. What is your accepted value and how does your accepted value compare with that of a second investigator?

At the end of the caliper opposite the jaws is found a depth gauge. The reading on the scale indicates the distance beyond the caliper which the gauge extends. The vernier is read the same as in the above exercises. Find an accepted value for the depth of a container (cylinder, beaker, etc.).

5. What is your accepted value and how does it compare with the accepted value of a second investigator?

Part B—The Micrometer Caliper

Examine a metric micrometer (Figure 2-1) caliper closely. Holding the instrument in the left hand, twist the thimble counterclockwise as in loosening a nut or a bolt. As the space between the anvil and spindle widens, the markings on the barrel are disclosed. Each mark represents one millimeter so that a reading correct to millimeters can be obtained directly from the barrel. Turn the thimble until any millimeter mark is showing when markings on the thimble are at zero. Now turn the thimble slowly in a counterclockwise direction until the next millimeter mark shows. Notice the thimble must turn twice between each mark. Since the thimble scale has 50 divisions and turns twice between each millimeter mark each division on the thimble represents 1/100 of a mm.

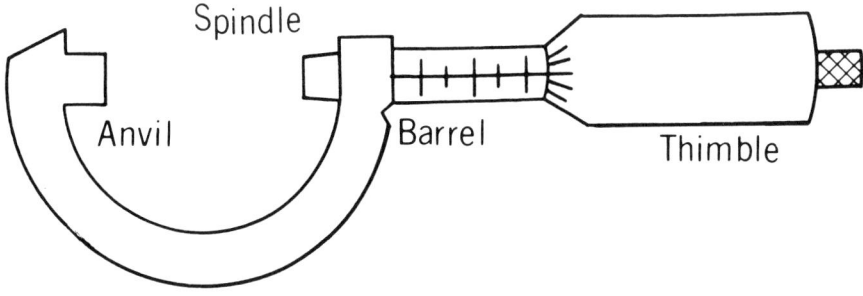

FIGURE 2-1

When each of three or four investigators feels he understands the procedures for using the micrometer caliper, let him find the thickness of some common metal substance such as a nail, a metal rod or metal sheet and record the results on a sheet of paper. *Do not permit the other investigators to see the results.* Now compare the results.
1. By how much do the results differ?
2. What factors might contribute to the differences?
3. Describe how one might obtain an accepted value for the thickness of the object.
4. Now each investigator should obtain *an accepted value* for the thickness of a human hair; a piece of fine copper wire.
5. Which of the two instruments used in this exercise yields measurements correct to the nearest .01 of a mm.?

DISCUSSION OF OUTCOMES

Part A

1. Individual results usually differ.
2. No result may be termed more correct than any other except for those measurements which are obviously different from the rest of the group.
3. The accepted value is obtained by finding the average of the values of the separate investigators.
4. Any reasonable value—accepted values should compare within several hundredths of a millimeter.
5. Any reasonable value. Accepted values should compare within several hundredths of a millimeter.

Part B

1. Results will usually differ by several hundredths of a millimeter.
2. Some factors that may be cited are: non-uniformities in the object being measured; differences in the micrometers; and misreading the micrometer.

Density of Solids 31

3. An accepted value may be obtained by finding the average of the values of several investigators.
4. This result will vary with each student and is designed merely to offer practice in the use of the micrometer.
5. The micrometer caliper.

EXERCISE 3—DENSITY OF SOLIDS

APPARATUS
1. Metric Rule
2. English Rule
3. Metric Weights
4. Avoirdupois Weights (if available)
5. Platform or Triple-beam Balance
6. Graduated Cylinder
7. String

(Teacher) An exercise concerned with the density of solids is included in the unit on measurement and forces for several reasons. The concept of density tends to elucidate and reinforce prior experiences with the metric system's weight-volume relationships. It helps to make clear the origin of the gram as a unit of mass and its relationship to the kilogram and liter. The second reason for including a density of solids exercise is that the procedure for determining the volume of an irregular solid by displacement of a liquid is an illustration of indirect measurement. The third reason is that density is an important concept not otherwise treated in this collection of exercises. The process skills likely to be strengthened by this exercise are measuring, experimenting and interpretation of data.

(Student) One very useful characteristic of a solid substance is its density. It should be remembered that mass units are free from any restrictions as to location while weight units are dependent upon the location of the substance. Since weight densities are commonly used, and vary only minutely on earth they are useful for many purposes. Density is defined as weight or mass per unit of volume. Since the density of any one substance is invariable and the likelihood of two different substances having the same density is very small, this physical characteristic may be used in making identifications and tests for purity.

PROCEDURE

Part A—Density by Measurement

Measure the length, width and height of a regular block in both the metric and English system. Make each measurement several times and determine an accepted value for each by finding the average of the trials.
1. Now find the volume of the block in both the metric and English systems.

Determine by weighing the weight of the the block in both grams and ounces, if possible.
2. Find the metric density. What units does it carry?
3. Find the English density. What units does it carry?

Part B—Density by Immersion

Some solids are irregularly shaped and their volumes cannot be computed by measurement. An irregular solid's density may be computed, however, after the volumes have been determined by the indirect method subsequently described.

Partly fill a metric graduated cylinder with water. Find and record the volume by reading the bottom of the curved surface of the liquid. (Note: If a graduated cylinder or measuring cup is available which bears English gradations the exercise may be repeated to find the density in this system.) Now lower an irregularly shaped object into the water until it is completely submerged. Record the new volume. On a platform balance weigh the solid and record the results.
1. What is the density of the irregular object?
2. If the density and volume of a material are known how might the weight be determined?

Part C—Teacher's Option

Find the density of a sheet of aluminum foil. The teacher will

Vector Forces 33

distribute pieces of foil approximately 3″ square. The student may find the volume of the foil by measurement with rule and micrometer, or by simple crumpling it and using the displacement method. If available balances are not sensitive enough to provide accurate measures of weight, refuse to provide larger pieces, and watch to see which student is first to borrow foil from others so that he may complete finding its density.

DISCUSSION OF OUTCOMES

Part A

1. The volume is computed by multiplying the length, width and height together.
2. Density is found by dividing the weight by the volume. The metric unit should be gram/cm^3.
3. Density is found as in answer 2. English density units are ounces/inch^3 or pounds/ft.^{3}.

Part B

1. Density will vary.
2. The weight of an object may be determined by multiplying its density times its volume.

EXERCISE 4—VECTOR FORCES

APPARATUS
1. Force board or table top
2. Three spring balances
3. String
4. Metric Rule
5. Paper

(Teacher) This exercise is designed to introduce the student to the concept of vector forces. The geometry of vector relationships is made understandable by means of the students' manipulations of small forces. Graphical methods are introduced as a tool for vector analysis. Process skills developed by this exercise are thought to be measuring, interpretation of data, experimenting, and formulating hypotheses.

(Student) A force is a *vector* quantity; that is it has both magnitude and direction. Force vectors are graphically represented by arrows. The length of the arrow represents the magnitude of the force and the direction of the arrow the direction of the force. When two or more forces act upon a common point, their combined effect, or *resultant force,* is dependent upon these two identifying properties of each force. It may also be shown that if the effects of two or more forces may be resolved into a single *resultant force,* then it is reasonable to expect that, with a force system in an unbalanced condition, a single force equal in magnitude to the resultant force, but opposite in direction, may be applied to produce a balance of forces called equilibrium. This balancing force is called an *equilibrant force.*

PROCEDURE

Part A—Examination of Multiple Forces

Attach two spring balances together with a loop of string and pull them in opposite directions.

1. How do the forces compare?

Now attach a third spring balance and apply an equal force to all three.

2. Under what geometric arrangement do the forces represented by the spring balance readings to appear to be equal?

Fasten one spring balance called A to a fixed object. Apply forces on the two remaining spring balances (B and C) so that they each read 500 g. when they are directed at an angle of approximately 90 degrees. (see Figure 4-1).

3. Is the sum of the readings on spring balances B and C equal to the reading of spring balance A?
4. What must the vector sum, or resultant, of forces B and C equal?

Vector Forces

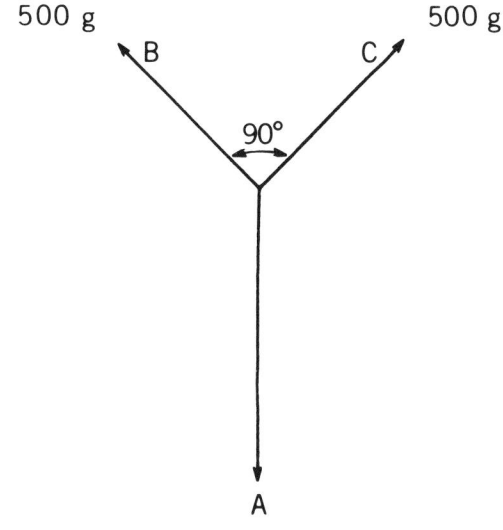

FIGURE 4-1

Part B—Graphical Analysis of Vector Forces

Attach one end of each of three spring balances together, apply force, and attach the other end of each balance to some firm object to establish an equilibrium of forces. The amount of force on each spring balance and the angle between them may be of any reasonable value. Now slide a piece of plain paper under the balances so that the center of the paper is at the intersection of the forces, and draw a line along each string. These lines represent the direction of each force. Label the lines A, B, and C in any order. Now extend the lines to their proper length by choosing an appropriate scale to represent each force as a vector. Using vectors A and B as adjacent sides, complete the parallelogram as shown in Figure 4-2. (Note: The length of sides A and B of the parallelogram must be *vector representations* of the forces A and B respectively.)

1. The reading on balance A is _____ g. and the vector representing this force is _____ cm. long.

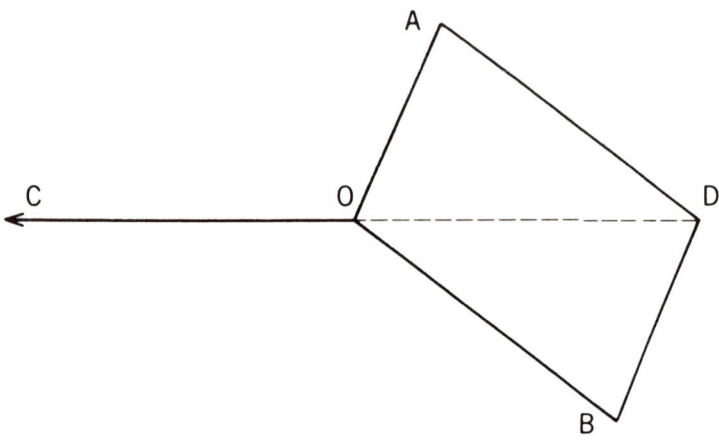

FIGURE 4-2

2. The reading on balance B is _____ g. and the vector representing this force is _____ cm. long.
3. The diagonal (OD) of the parallelogram is _____ cm. long and therefore represents a force of _____ g.
4. The reading on spring balance C is _____ g. How does this compare with the force represented by OD? _____.
5. The diagonal OD is the _____ of the two forces A and B. Since this force is approximately equal in magnitude but opposite in direction to force C, force C must then be called a(n) _____.

Part C—Testing a Graphical Solution

Draw on a piece of plain paper a diagram representing two forces acting upon a single point. Graphically determine the magnitude and direction of an equilibrant force that might be applied to the system to produce equilibrium. Test this vector diagram by duplicating the vector system and determining the magnitude of the force with spring balances.

Resolution of Forces

DISCUSSION OF OUTCOMES

Part A

1. The forces are equal.
2. The forces on the spring balance will appear to be equal when each is separated from the others by an angle of 120°.
3. The sum of the readings on spring balances B and C does not equal the reading on spring balance A.
4. The vector sum of forces B and C is approximately 700 g. and is equal to force A.

Part B

1. Any reasonable value.
2. Any reasonable value.
3. Length to be taken from the student diagram.
4. The force on spring balance C should equal the force represented by diagonal OD.
5. Resultant; Equilibrant force.

Part C

A good method to recommend to the students for testing a vector diagram is to lay the vector diagram upon a table top and use the diagram itself to provide direction for the forces. All that remains of the test is to produce forces equal to, or in the same ratio as, the forces on the vector diagram.

EXERCISE 5—RESOLUTION OF FORCES

APPARATUS

1. Three spring balances
2. Upright Support
3. Boom
4. String
5. 500 g. weight
6. Plain Paper

38 Measurement and Forces

(Teacher) Exercise 4 instructs students in the process of combining two or more forces in order to find the magnitude and direction of a resultant. This exercise, Exercise 5, provides a graphical method whereby the student may resolve a single force into two or more component forces. This sort of analysis is useful in examining stresses set up in a structure by a single applied force. Along with introducing the student to a system of vector analysis by the graphical method, this exercise is thought to aid in the development of the process skills of measuring, interpretation of data and experimenting.

(Student) It is sometimes useful to consider a single force as being made up of two or more forces. The force exerted on the handle of a lawn mower, for instance, may be thought to have a horizontal component which causes the mower to move forward and a downward component which presses it to the ground. This process is said to be resolving a force into its *component* forces, or simply the *resolution of forces*. The magnitude of the components may be determined by considering the force vector to be the diagonal of a parallelogram and the components to be the sides.

PROCEDURE

Part A

To graphically resolve a force of 200 g. into two components acting at angles of 30° and 60° to the direction of the force, and at right angles to each other, first, in the approximate center of a plain piece of paper and using a scale 1 cm. = 20 g., draw a vector AB representing the 20 g. force. Letting AB be the diagonal of a parallelogram, construct sides AC and AD as shown in Figure 5-1 at angles of 30° and 60° respectively to line AB. A line perpendicular to AD and intersecting point B, and a line perpendicular to AC and intersecting point B will complete the parallelogram.

1. The length of AD is _____ cm. and represents a force of _____ g.
2. The length of AC is _____ cm. and represents a force of _____ g.

The results of this graphical solution may be substantiated

Resolution of Forces

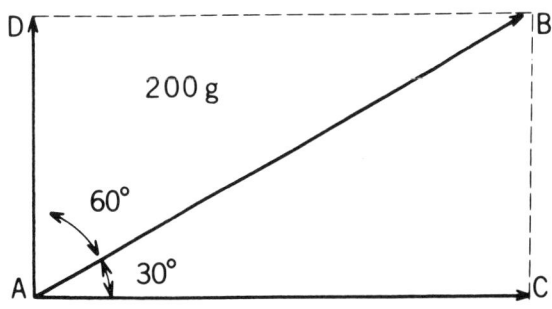

FIGURE 5-1

experimentally with the following procedure: Attach one end of a spring balance to some fixed object. To the other end attach two additional spring balances. With one partner applying a force equal to vector AC at an angle of 60° to the axis of spring balance A, and a second partner applying a force equal to vector AB at angle of 30° to the same axis (as shown in Figure 5-2), read and record the force exerted on spring balance A.

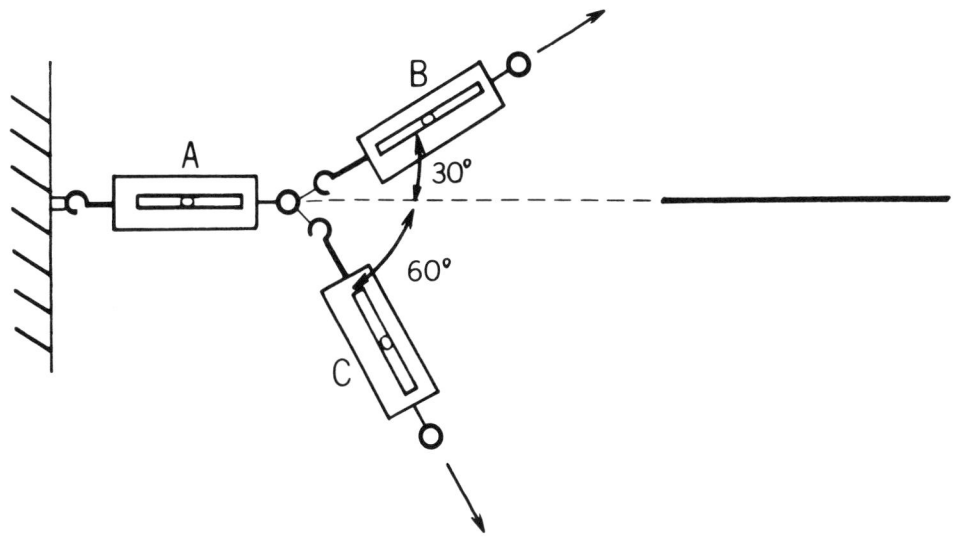

FIGURE 5-2

3. The reading on spring balance A is _____ g.
4. How does this reading compare to a 200 g. force?

5. Find the percent of error accepting 200 g. as the correct reading.
6. Considering the intruments employed and the method of experimentation, do you consider the evidence sufficient to accept the method of graphical analysis of vectors?

Part B

In Figure 5-3 the guide wire and the boom are supporting the 500 g. weight, and the vector sum of the forces directed along these two paths is therefore 500 g. Set up the apparatus as shown in Figure 5-3, and with the boom as close to horizontal as possible, record the force exerted upon the guide wire by noting the reading

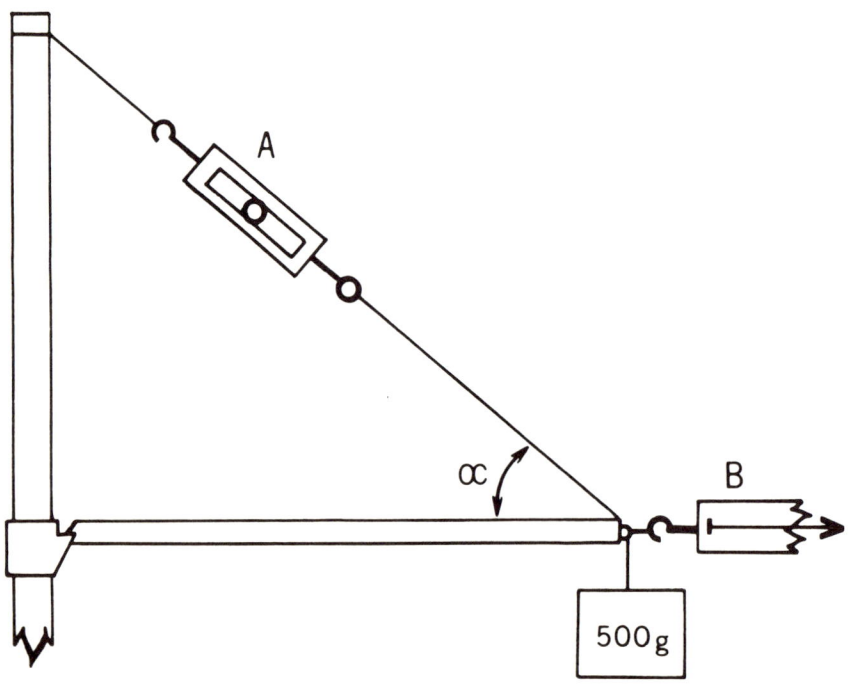

FIGURE 5-3

Resolution of Forces

on spring balance A. (Be sure to subtract out any part of the reading that represents the weight of the spring balance.)

1. The tension on the guide wire is ———— g.

Now pulling horizontally on spring balance B with just enough force to overcome the compressive force on the boom note the reading on the balance.

2. The horizontal force exerted on the boom is ———— g.
3. The angle (α) enclosed by the guide wire and boom is ———— degrees.
4. Can you write a statement regarding the relationships between a force on an object caused by gravity and the forces required for its support?

For comparison the components will also be determined graphically. On a piece of plain paper draw a vector AB representing the force of 500 g. Draw line DAC perpendicular to the vector and passing through the tail. Now construct line EA so that it forms the angle with line DAC. To complete the parallelogram, draw line EB parallel to DAC and passing through point B, and line BC parallel to EA and passing through point B (see Figure 5-4).

5. By measurement the length of AC is ———— cm.

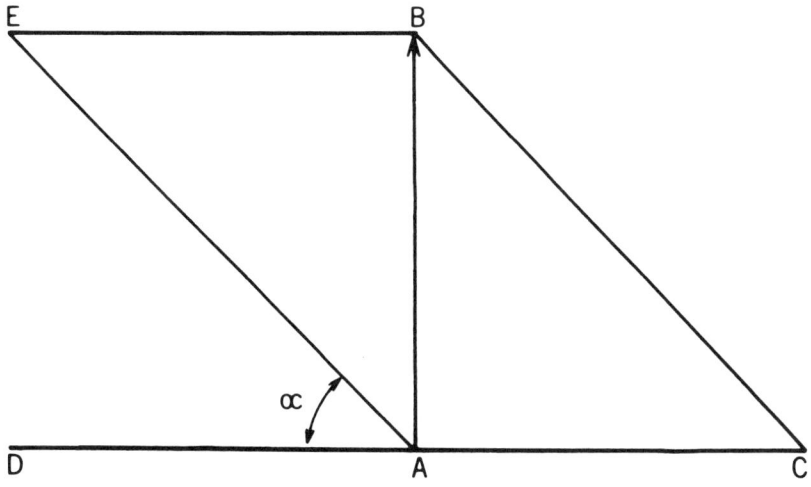

FIGURE 5-4

and represents a force of _____ g. How does this compare with the force on spring balance B above?

6. By measurement the length of BC is _____ cm. and represents a force of _____ g. How does this compare with the force on spring balance A above?

7. Has the graphical method proved to be a satisfactory method of determining component forces?

DISCUSSION OF OUTCOMES

Part A

1. 5 cm. (approx.), 100 g.
2. 8.7 cm. (approx.), 174 g.
3. The reading on spring balance A should approximate 200 g.
4. Student answer. Because of the technique and normal inaccuracies of spring balances the discrepancy between the expected and attained readings may vary 20 to 30 grams.
5. The percent of error =
$$\frac{\text{the difference between 200 g. and reading A}}{200}$$
6. Student answer. This is a good time to discuss with the students the point that the degree of accuracy an invesigator might expect from an experimental situation is is dependent upon the sophistication of the instruments and methodology employed. Error if rationally accounted for does not always preclude the drawing of conclusions from experimental evidence.

Part B

1. Student answer—the magnitude of which is dependent upon angle alpha, but should always be greater than 500 g.

2. Student answer the magnitude of which is dependent upon angle alpha, but should always be less than force A.
3. Student answer from direct measurement.
4. Student statement, the gist of which should be that the forces which are required to support an object may exceed the weight of the object.
5. Student answer. The result of the graphical solution should compare favorably with the spring balance reading.
6. Student answer. The result of the graphical solution should compare favorably with the spring balance reading.
7. Student conclusion. The teacher must help students to properly rationalize this statement for they often are too demanding of accuracy in their data in relationship to the nature of the exercise.

EXERCISE 6—THE LAW OF MOMENTS

APPARATUS
1. Meter stick
2. Metric weights
3. String
4. Triangular wood block or rigid upright support

(Teacher) The law of moments is an important physical concept which is significant in its direct application to technology. This exercise attempts to illuminate this concept for the student and at the same time provide him with added practice in making measurements and doing calculations in the metric system. The process skills employed in the completion of the exercise are measuring, and in Part C hypothesizing, controlling variables and experimenting.

(Student) A rigid bar which is free to rotate about a fixed point, or *fulcrum,* is a *lever.* The tendency of the lever to rotate when a force is applied is called *torque.* Torque is equal to the force applied times the length of the *moment arm* of the force, the *moment arm* being defined as the perpendicular distance between the point where the force is applied and the fulcrum. When a lever does not rotate about the fulcrum a state of *equilibrium* exists wherein the sum of the torques, or *moments,* which tend to produce rotation in a *clockwise* direction, equals the sum of the torques, or *moments,* which tend to produce *counterclockwise* rotation.

PROCEDURE

Part A—Checking the Law of Moments

The law of moments states that when a lever is in equilibrium the sum of the clockwise moments equals the sum of the counterclockwise moments. Balance a meter stick on edge upon a triangular block, or by hanging it from a rigid support. (*Note*: Some meter sticks are of non-uniform density and will not balance at the 50 cm. mark.)

When the meter stick is balanced and is in a state of equilibrium the weight of the meter stick may be hereafter ignored when considering unbalanced forces.

Now suspend by a string a 200 g. weight which is located 10 cm. to the left of the fulcrum. On the other side of the fulcrum slide a 100 g. weight along the meter stick until the system is again in equilibrium (see Figure 6-1).

1. What is the moment arm of the 100 g. weight?
2. The _____ weight tends to produce counterclockwise motion, while the _____ weight tends to produce clockwise motion.

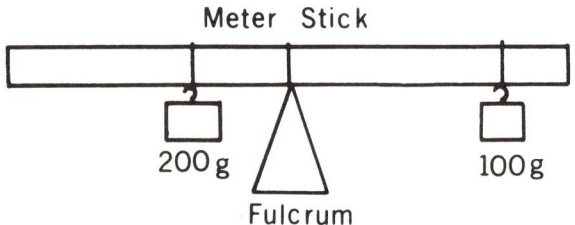

FIGURE 6-1

The Law of Moments

3. What is the clockwise torque? Remember torque equals the product of an applied force times its moment arm.
4. What is the counterclockwise torque?
5. Is there a significant difference between the clockwise and counterclockwise torque?

Part B—Multiple Forces and the Law of Moments

To test the law of moments when more than one force is applied in the same direction, reestablish the equilibrium of the apparatus as in Part A. Suspend a 100 g. weight on the same side of the fulcrum as the 200 g. weight. Now adjust a 50 g. weight on the opposite side of the fulcrum until equilibrium is again attained (see Figure 6-2).

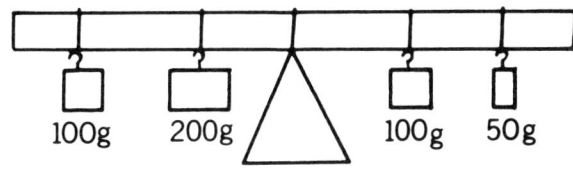

FIGURE 6-2

The total clockwise torque is the sum of the two separate torques produced by the weights to the right of the fulcrum.
1. What is the total clockwise torque?
2. What is the total counterclockwise torque?
3. Is there a significant difference between the clockwise and counterclockwise torque?

Part C

Reestablish the equilibrium described in Part A. If a weight were added to the system in such a way as to add 12,500 g-cm. of torque to the torque produced by the 200 g. weight, where would a 1000 g. weight be placed to produce equilibrium?
Test your hypothesis.

DISCUSSION OF OUTCOMES

Part A

1. 20 cm.
2. 200 g., 100 g.
3. 2000 g-cm. Torque equals 100 (20).
4. 2000 g-cm. Torque equals 200 (10).
5. There should be no significant difference between the two torques.

Part B

1. Student answer. Clockwise torque equals 100 (20) plus 50 (Ma.).
2. Student answer. Counterclockwise torque equals 200 (10) + 100 (Ma.).
3. There should be no significant difference between the two torques.

Part C

The 1000 g. weight should be placed on the same side as the 100 g. weight at a distance of 12.5 cm. from the fulcrum.

EXERCISE 7—A NON-UNIFORM LEVER

APPARATUS

1. Meter stick
2. Clamp or weight hanger
3. Metric weights

(Teacher) An exercise dealing with the non-uniform lever provides the student with some insight into a technological application of the law of moments. Also, the concept of center of gravity is clarified. The activity is designed to provide the student further practice in the use of the metric system and provides a problem designed to test his ability to apply the principle with which he has been working.

A Non-Uniform Lever

(Student) The weight of a uniform lever, such as a meter stick, is distributed fairly evenly along its entire length. When a fulcrum is placed at the midpoint of the bar, there is an equal amount of weight on each side of the fulcrum, and therefore a state of equilibrium exists.

The center of the bar is then said to be the center of gravity; that is the point where all the weight of an object may be considered to be concentrated. A non-uniform lever will also have a center of gravity. It will not be located halfway along its length, but somewhere closer to the heavier end of the bar. It is at this point all the weight of the lever is considered to be concentrated. When the fulcrum is placed other than at the center of gravity an equilibrium does not exist and torque must be added to create equilibrium.

PROCEDURE

Part A

To create a non-uniform lever fasten a clamp or weight hanger to one end of a meter stick. Now find the center of gravity.

1. How did you do it?

Now move the fulcrum 10 cm. in a direction away from the weighted end of the meter stick and attach a 200 g. weight to the lighter end, adjusting the distance until a state of equilibrium is achieved (see Figure 7-1).

2. What is the moment arm of the torque created by the weight of the lever.

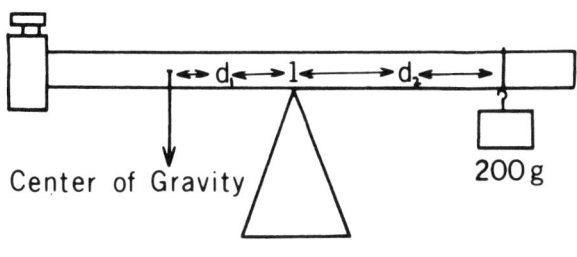

FIGURE 7-1

48   *Measurement and Forces*

3. What is the moment arm of the torque created by the 200 g. weight?
4. Using the law of moments, solve for the weight of the lever.

Note: If the student is not familiar with the law of moments, he may refer to Exercise 6 for information.

Part B

Repeat Part A moving the fulcrum 15 cm. from the center gravity and using 100 g. as a counter balancing weight.
1. What is the moment arm of the torque created by the weight of the lever?
2. What is the moment arm of the torque created by the 100 g. weight?
3. Can you determine the weight of the lever?
4. How does this result compare with that of Part A?
5. Compute the percent of difference using the result of Part A as the accepted value.
6. Test your ability to apply what you have learned about non-uniform levers on the following problem.

A tapered pole 20 ft. long has its center of gravity 5 ft. from the heavy end. If a fulcrum is placed at the center of the pole's length how much force is necessary to lift the heavy end off the ground if the pole weighs 300 pounds?

DISCUSSION OF OUTCOMES

Part A

1. The simplest procedure to determine the C of G is to find the point on the lever at which it balances.
2. Student answer.
3. Student answer.
4. The weight of the lever is determined with the following relationship: $(W)(10) = 200 (d_2)$.

Part B

1. Student answer.

2. Student answer.
3. The weight of the lever is determined with the following relationship: W (15) = 100 (d_2).
4. The weights of the lever found by two sets of experimental conditions should compare favorably.
5. Any reasonable percent, usually less than 3%.
6. 225 lb.; 20 F = 15 (300).

EXERCISE 8—PARALLEL FORCES

APPARATUS

1. Meter stick
2. 2 Spring balances (0-250 g)
3. String
4. Weights
5. Platform Balance

(Teacher) The concept of parallel forces has many practical applications in engineering and technology. This exercise is thought to provide the students with practice in the use of the metric system. It also requires of them the skills of drawing generalizations and making applications of previously learned concepts.

(Student) Parallel forces act parallel to one another and may be in the same or opposite directions. There are many practical instances of parallel forces acting in the same direction. Some examples would be: piers supporting a bridge, ladders supporting a scaffold, the shoulders of two boys supporting a pole between them, etc. For the purpose of determining the magnitude of forces acting upon such a system it is convenient to select the point at which one force acts to be considered as a fulcrum. In Figure 8-1, for instance, A is selected as the fulcrum and the torque which results from the force at A is zero, for it has no moment arm. The clockwise torque is the result of the force (W) times its moment arm (D_1). The counterclockwise torque is the result of the force at (B) times its moment arm (D_2). Since a state of equilibrium exists, clockwise torque = counterclockwise torque. $W \times D_1 =$ Force B $\times D_2$.

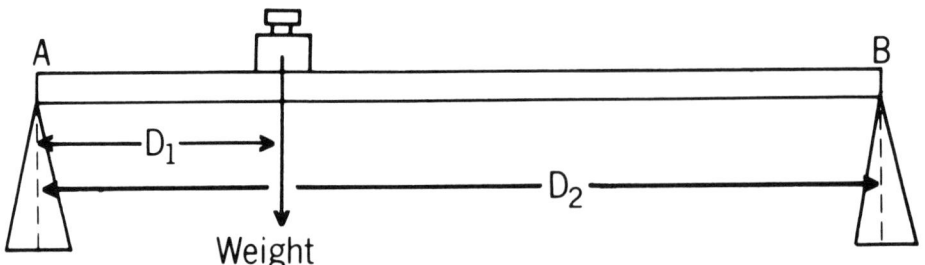

FIGURE 8-1

PROCEDURE

Part A

Determine the weight of a meter stick on a platform balance. Suspend the meter stick, as in Figure 8-2, 10 cm. from each end by attaching them to spring balances. Now record the reading of each spring balance.
1. How does the sum of the spring balance readings compare with the weight of the meter stick?
2. Considering the nature of the activity and the apparatus, does this evidence support the principle that the sum of the upward forces equals the sum of the downward forces?

Now suspend a 500 g. weight 25 cm. to the left of the center of gravity and record the readings of the spring balances. (Note: The center of gravity of the meter stick will usually coincide with the 50 cm. mark.) Now record the lengths of the moment arms *considering A to be the fulcrum.*
3. What is the clockwise torque of the system?

The clockwise torque is the sum of the torques created by the 500 g. force and the weight (w) of the stick. (Clockwise torque = $Ma_1 \times 500$ g. plus $Ma_2 \times$ Weight of the meter stick.)

The counterclockwise torque is represented by the force at B times its moment arm. Since clockwise torque equals counterclockwise torque, then F_B may be determined as follows: Clockwise torque (from item 3) equals $F_B \times$ (moment arm).
4. What is the computed force F_B at B?

Parallel Forces

5. What is the reading of spring balance B?
6. Find the percent of difference between the reading of spring balance B and the computed F_B.
7. If the percent of difference found in item 6 is small this is evidence which supports what law?

Part B

Now let B be considered the fulcrum and determine experimentally the force at A. Compare the computed force at A with the reading on spring balance A.

Part C

If an investigator knows the total weight involved in a system, such as is exhibited in Figure 8-2, and also knows the force exerted at one of the supports, how may he readily find the force at the other?

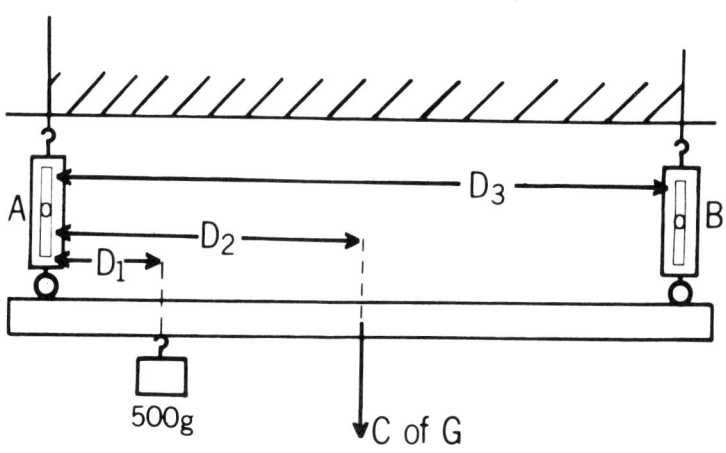

FIGURE 8-2

DISCUSSION OF OUTCOMES

Part A

1. The sum of the two spring balance readings should compare favorably with the weight of the meter stick.

2. If item 1 is answered in a positive way then the generalization called for is warranted.
3. Any reasonable answer. Commonly students will neglect the fact that the meter stick is suspended at the 10 cm. and the 90 cm. marks causing them to employ erroneous moment arms in their computations. The equation should be: Clockwise torque = 15 (ma) × 500 (f) plus (ma) × Weight of meter stick.
4. Any reasonable answer. The same caution concerning moment arms as presented in item 3 applies here. Answers will usually range from 140 to 180 g-cm.
5. Student answer.
6. The computed force at B and the reading on spring balance B should coincide reasonably well.
7. The Law of Moments.

Part B

The force at A with B as the fulcrum is found with the following equation: $F_A \times 80 = 500 (65) + 40 \times$ (weight of the meter stick). The computed force should compare favorably with the spring balance reading at A.

Part C

Since in an equilibrium condition the sum of the upward forces must equal the sum of the downward forces the second upward force may be determined by subtracting the known upward force from the total downward force.

EXERCISE 9
OPTIONAL PROBLEMS TO TEST LABORATORY SKILLS

APPARATUS

Materials and apparatus for this exercise include any that are found in the school, home or laboratory. The teacher may advise students as to the availability of equipment, but should not attempt to supply all that is requested by the students.

Optional Problems

(Teacher) This is the first of the exercises which appear at the conclusion of various laboratory units that require the students to assume some measure of responsibility for experimental design. The problems which the students are asked to investigate to conclude Unit I are relatively simple since they represent the first efforts of most of the students at independent investigation. Also at this point in their training they have been exposed to relatively few experimental designs and have likely acquired only a limited number of manipulative and process skills.

Students will require varying amounts of assistance and prodding. Care must be taken to provide enough direction to keep the investigations moving toward solutions, but not provide so much information that the students are relieved of the primary responsibility for the design and execution of their studies. The teacher's function becomes one of advisor and consultant to the many independently operating student investigators.

Though the situation will vary with the group, it has been found that the optimum number of students working on an investigating team is three to four. It appears the idea pool of several students offers some advantage in terms of time-saving over the lone operator. More than four students in each group appears to add to the number that operate only as spectators.

The time required for this exercise *may* run to four laboratory periods. Sometimes in addition to this allotted time, some students will need to find time after school and during study periods to complete the project. Most students in this program have accepted the challenge and put in the extra time on a voluntary basis. The enthusiasm of the students for physics investigation is one of the most positive outcomes of the program.

(Students) With your present knowledge of laboratory procedures and skills in handling equipment it is time to accept the challenge of devising and putting into action your personal program for the study of a scientific relationship. Two separate problems, along with suggestions for their study, will be presented. It is up to you to devise and carry out plans for the study and to draw reasonable conclusions from the data.

In writing the laboratory report, be sure to include the following:

(1) An introduction which includes a statement of the problem and a general description of proposed procedures.
(2) A fairly detailed description of the procedures including the data obtained. This log should contain a record of *all* data and activities, including any which proved to be of little value in the solution of the problem.
(3) A summary and statement of conclusions.

Problem A—Density of a Fine Wire

(Teacher) Each investigating team may be issued a short length (10-12") of number 20 or 22 copper wire. Essentially the students will need to establish the weight and volume of the wire in order to establish the density. They have been introduced to micrometers and balances which are the necessary tools for dealing with the problem. If you wish to challenge the students' ingenuity, you may put "off limits" any microbalances or small weights which allow for the direct determination of the weight of the wire. The students would then be required to invent some device which employs the law of moments in order to determine the weight of the wire specimen.

(Student) You have been given a specimen of copper wire. Your job is to determine its density. Review the exercises which you have performed in the past laboratory periods and devise a method for doing this. Each laboratory team should work independently and devise its own procedure for solution of the problem. This is your chance to show that you are capable of operating independently and without directions. It is your chance to show that you are capable of formulating your own procedures for investigation of a problem. Can you meet this challenge?

Problem B—Bridge Building

(Teacher) In this exercise the students are asked to apply their understanding of the resolution of forces to a practical problem. Each team of investigators is asked to design and build the strongest possible 6 inch long bridge from a specified list of supplies. The supplies may include: 15 (or some other number)

Optional Problems

soda straws, 2 wooden blocks (upon which the bridge rests), 10 ft. of cotton twine, a tube of airplane glue, thumb tacks, and a board 2 ft. long upon which the whole structure rests. (Note: The 2 ft. board may also serve as the anchor for the system.) The design of the bridge must be such that its strength may be tested by the addition of sufficient laboratory weights to cause its collapse. Each team must present a diagram of the distribution of the stresses for their design and their reasons for choosing the design.

It is often an entertaining and informative procedure to have each group formally present their structure and their rationale to the class. The strength of the presenting group's structure is then publicly tested by adding the amount of weight necessary to overload it and to cause it to collapse.

(Student) The test of your understanding of force vectors is your ability to apply some of the concepts to a practical problem. The problem in this instance is to build the strongest possible structure to bridge a gap between two wooden blocks. The blocks are to be six inches apart and may rest on a larger board which will serve as a base for the project. You may use only those supplies designated by your teacher.

Your finished project will be presented to the class along with a force diagram and an explanation of the distribution of the stresses in the structure. The teacher may choose to test your bridge by adding sufficient weight to cause its collapse. Therefore the design must be such as to allow the placement of laboratory weights upon it for such a test.

Supplies which may be allowed in the construction of the bridge are: one board 2 ft. in length, two 2" x 4" wooden blocks, 15 paper soda straws, 10 ft. of cotton twine, airplane glue, thumb tacks, and any other materials approved by the teacher.

This is your chance to show your ability as an independent investigator, engineer, and builder. Let us see what kind of job you are capable of doing.

Unit II

Mechanics of Solids

The Pendulum
Pulley Systems
Friction
The Inclined Plane
Hooke's Law
The Wheel and Axle
Velocity and Acceleration
Conservation of Momentum
Optional Problems to Test
    Laboratory Skills

Unit II

EXERCISE 10—THE PENDULUM

APPARATUS
1. Wood Bob
2. Metal Bob
3. Light String
4. Meter Stick
5. Timer
6. Protractor

(Teacher) The pendulum is a simple device that may be adapted quite readily to exercises which stress the process skills. In this exercise, the students are required to make operational definitions, formulate hypotheses concerning the variables which affect the period of the pendulum, control and manipulate variables, and experiment. In fact, this activity so well demonstrates the process skills many teachers may wish to adapt it for a classroom demonstration.

To use the exercise as a classroom demonstration, the teacher would:

1) Ask the class for terms which define the various positions, motions, and time relationships of the pendulum. These terms would be related to position, period, frequency, amplitude, cycle and time. Student terminology for these relationships should be accepted.

2) Ask the class for hypotheses as to variables which will affect the period or the number of cycles in a time period, say twenty seconds. The hypotheses should be complete statements such as: A heavier weight will result in a decrease in the number of oscillations occurring in twenty seconds. Most classes will hypothesize about the weight of the bob, the length of the string, and the amplitude of the arc of the pendulum.

3) At this point, the teacher may turn the testing of the hypotheses over to the students asking them separately, or in small groups to find out whether their hypotheses are correct. Of course, the teacher may also wish to complete the hypothesis testing as a demonstration. It will be found that the only variable which affects the period of the pendulum is its length. (Position as related to earth is neglected for obvious reasons.)

PROCEDURE

Part A

(Student) You are about to perform an exercise which will determine your ability to formulate the test hypotheses. The vehicle for exercising these skills is the pendulum. A simple pendulum may be made by suspending a wooden bob with a piece of string from a rigid support. Construct a pendulum and watch it swing to and fro. The *period* of the pendulum is the time required for one *cycle* from A to C and back to A (see Figure 10-1).

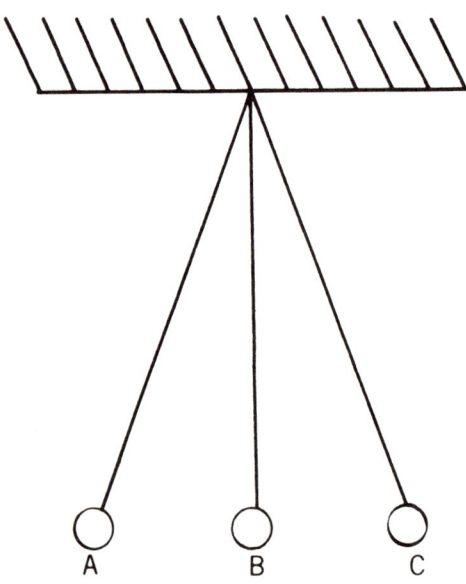

FIGURE 10-1

The Pendulum

The *amplitude* is the distance the bob swings from its rest position, and the *frequency* is its number of cycles per second.

(Teacher) You may wish to use the terminology as devised by the class. Here are two questions which you might ask:
1. How many cycles does your pendulum complete in one minute?
2. What factors (variables) do you think affect the number of cycles the pendulum completes in one minute?

Write out complete statements as to how each variable will affect the number of cycles/minute of the pendulum. These become your working hypotheses. (Teacher: You may wish to check each laboratory group's hypotheses to make sure they include weight of bob, length of string, and amplitude of arc.) Now test each hypothesis you have formulated after carefully considering which variables must remain constant and which are to be adjusted. Remember only one hypothesis can be tested at a time. All variables except that one dealing with the hypothesis being tested should remain constant. A written report should be made stating each hypothesis, the procedure used for testing the hypothesis, and your conclusion as to whether the hypothesis should be accepted or rejected. Note to Students: Amplitudes of arc should be kept within a 20° maximum for consistent results.

(Teacher) You may wish to assign the parts B and C of the exercise to those students who have completed Part A.

Part B—Teacher's Option

(Student) You have discovered that the length of a pendulum affects the number of cycles it makes in a given period of time. Can you discover the exact relationship between the number of cycles/minute and the length of the pendulum? (Hint: Increase the length by four times and by nine times.)

Part C—Teacher's Option

The *frequency* of a pendulum is defined as its number of cycles per second:

$$\frac{\text{cycles/min.}}{60 \text{ sec./min.}}$$

The *period* of a pendulum is defined as the time in seconds required for one complete cycle and is the reciprocal of the frequency. The period (P) of a pendulum has been found to be related to its length (L) and the acceleration constant of gravity (g) in the following way: $P = 2\pi \sqrt{1/g}$. Can you find (g) for your location on earth? You may use data previously derived in this exercise.

DISCUSSION OF OUTCOMES

Part A

1. Student answer varies with the length of each pendulum.
2. Student answers should include weight of the bob, length of string, amplitude or arc, and any others they may deem appropriate.

Part B

Students should find that increasing the length by a factor of 4 will increase the period by a factor of 2 and increasing the length by a factor of nine will increase the period by a factor of three. The period is found to be related to the length by: $\dfrac{P_1}{P_2} = \sqrt{\dfrac{l_1}{l_2}}$

Part C

Solving the relationship $P = 2\pi \sqrt{1/g}$ for g yields: $g = \dfrac{4\pi^2 1}{P^2}$ when the angle of oscillation is small. By substitution the students should obtain results which approximate 9.8 meters/sec.^{2} or 32 ft./sec.^{2}

EXERCISE 11—PULLEY SYSTEMS

APPARATUS

1. String
2. Two single pulleys
3. Two double pulleys
4. Weights
5. Spring balance
6. Platform balance

Pulley Systems

(Teacher) An exercise dealing with pulley systems may benefit students in several ways. They may come to understand that technology is the direct application of empirical law, and engineering often depends upon simple relationships which have been established by testing of apparatus and materials in physical testing laboratories. A knowledge of simple machines may, through a small number of exercises such as this, increase their appreciation of the technological aspects of science. In this somewhat traditional exercise, the students are exposed to the process skills of observing, measuring, and inferring. In Part E (optional) of the exercise, predicting and experimenting skills are utilized. The predictions the students are asked to make in this section resemble very closely the process of formulating deductive hypotheses. That is, they are called upon to infer and predict relationships based upon prior knowledge of similar relationships.

(Student) A *pulley* is a simple machine consisting of a grooved wheel free to revolve about an axle. Pulley wheels are incorporated into a single sheath or block in many different combinations for various purposes. A reason for the use of a pulley may be to enable man to lift more than the ordinary weights or merely to change directions of the force used to overcome a resistance. The ratio of resistance overcome to input force is the *Actual Mechanical Advantage* of any machine:

$$\text{A.M.A.} = \frac{\text{Resistance Overcome}}{\text{Effort Force}}$$

The *Ideal Mechanical Advantage* of a pulley system may be readily determined by counting the number of strands that actually support the resistance.

PROCEDURE

Part A—A Single Fixed Pulley

Set up the single fixed pulley as shown in Figure 11-1. Raise, then lower the 500 g. weight *at a uniform rate,* recording the reading of the spring balance as it moves each way. Do three trials of

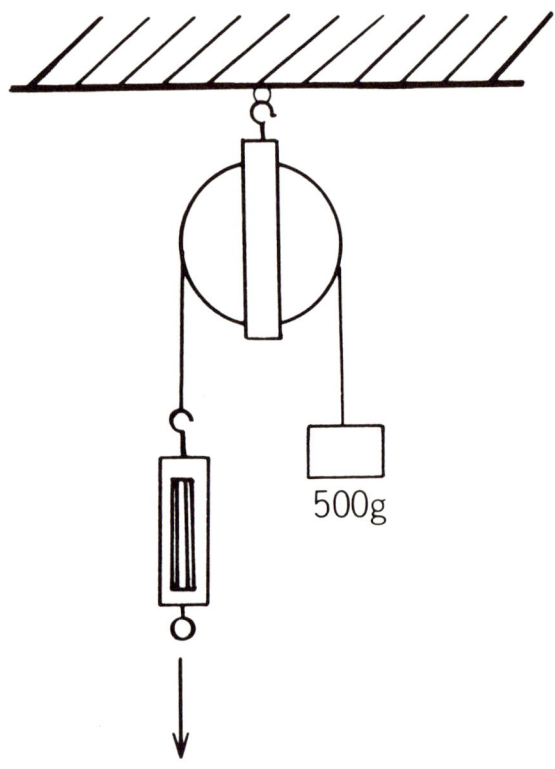

FIGURE 11-1

each. (Note: Be sure to hang the spring balance so that it records its own weight.)

1. What is the average of the forces both ways?

This represents the force necessary to overcome 500 g. of resistance with a single fixed pulley.

2. If M.A. = $\dfrac{\text{Resistance Overcome}}{\text{Effort Force}}$, what is the mechanical advantage of a single fixed pulley?

3. How many strands of string actually support the weight?

Part B—A Single Movable Pulley

Set up the single movable pulley as shown in Figure 11-2. Raise and lower the 500 g. weight at a uniform rate and record the spring balance reading. Weigh the pulley.

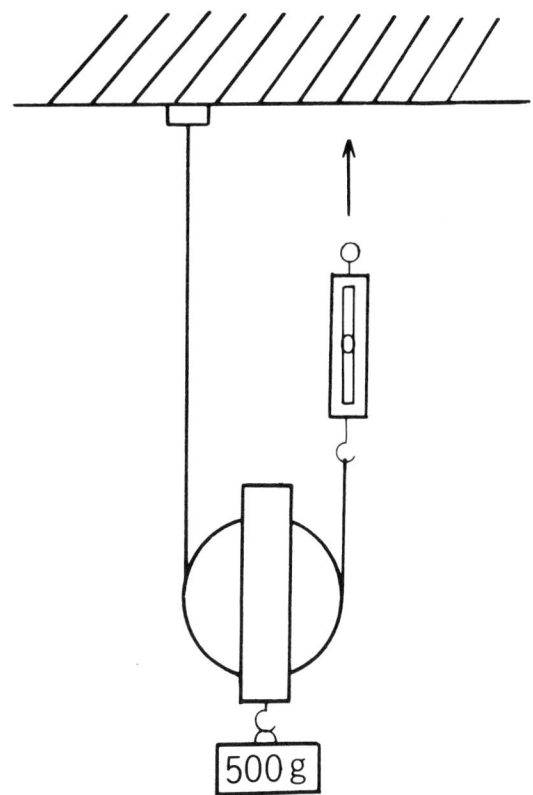

FIGURE 11-2

1. The total resistance must be considered to be the sum of what two forces?
2. After taking spring balance readings for three trials, find the average of forces both ways.

3. What is the mechanical advantage of this pulley system?
4. How many strands actually support the resistance?

Part C—One Fixed, One Movable Pulley

After weighing the single, movable pulley, set up the systems as shown in Figure 11-3.

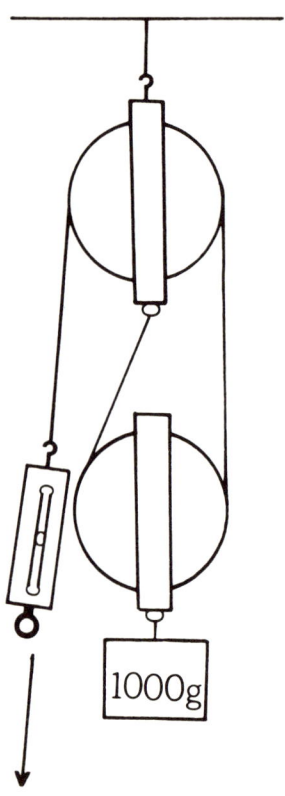

FIGURE 11-3

1. The resistance in this system is comprised of what two forces?
2. Determine the approximate effort necessary to overcome

Pulley Systems

the resistance in this system by finding the average of six trials, three moving the weight down, and three trials moving the weight up.
3. What is the mechanical advantage of the system?
4. How many strands actually support the resistance?

Part D—Two Fixed, Two Movable Pulleys

After weighing the double movable pulley, set up the system as shown in Figure 11-4.

1. The resistance in this system is the result of what two forces?
2. Find the approximate effort necessary to overcome the resistance in this system by finding the average of six trials, three moving the weight up and three moving the weight down.
3. What is the mechanical advantage of the system?
4. How many strands actually support the resistance?

Measure the distance the effort moves as the resistance is raised 10 cm.

5. If work = force × distance, how does the work output (movement of the resistance) of this pulley system compare to the work input (movement of the effort)?
6. How does the increase in potential energy (P.E. = Resistance × increase in height) of the system compare with the work input?
7. Go back to the data gathered in Parts A through D which concerns the mechanical advantage of each system and the number of supporting strands. Can you state the relationship that exists between these two factors?

Part E—Teacher's Option

Predict the effort necessary to raise a textbook (or other object) at a uniform rate of speed with each of the pulley systems. Test your predictions. Write a description of the procedures and results of Part E. Account for any discrepancies in your results.

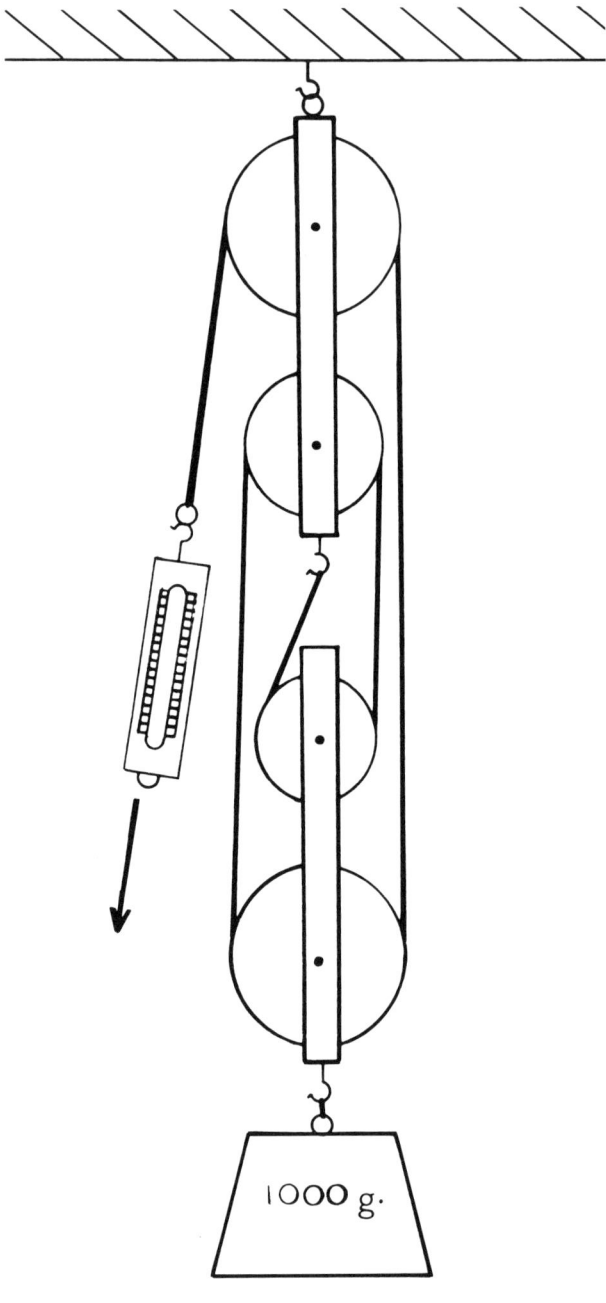

FIGURE 11-4

Pulley Systems 69

DISCUSSION OF OUTCOMES

Part A

1. Approximately 500 g.
2. One to one
3. One

Part B

1. The resistance is the sum of the 500 g. weight and the weight of the pulley.
2. The spring balance reading should approximate one-half the total force represented by the sum of the 500 g. weight and the weight of the spring balance.
3. The mechanical advantage is approximately 2 to 1.
4. Two strands actually support the resistance.

Part C

1. The resistance is the sum of the 500 g. weight and the weight of the pulley.
2. The effort should approximate one-half of the magnitude of the resistance.
3. The mechanical advantage is two to one.
4. Two strands actually support the resistance.

Part D

1. The resistance is the sum of the 1000 grams of weight and the weight of the movable pulley.
2. The effort should equal one-fourth the magnitude of the resistance.
3. The mechanical advantage is 4 to 1.
4. Four strands support the resistance.
5. Theoretically work output of the system should equal work input. However, due to friction the work output will be slightly less than the the work input.
6. The increase in potential energy (P.E. = Resistance × increase in height) should approximately equal the work done on the system.

7. The ideal mechanical advantage of a pulley system is equal to the number of supporting strands.

EXERCISE 12—FRICTION

APPARATUS

1. Wood block
2. Weights
3. Spring Balance

(Teacher) An exercise dealing with friction, besides acquainting the student with some properties of this common and very important concept, adapts itself to laboratory experiences that may reinforce several of the process skills. The student is asked to make predictions and hypotheses and to test them. At least one of the exercises, the effect of surface area upon friction, normally produces a discrepancy between student prediction and experimental result. Here the student is faced with a dilemma. Was there an error in his experimental procedures or is the relationship different than he first surmised? The way he meets this challenge is very basic to his understanding of the scientific method. Data interpretation skills of the student are tested and strengthened at the point where the student must decide if the coefficient of friction is a constant for the particular block of wood and table surface used in the exercise. In this exercise, process skills may be emphasized, for the basic relationships dealing with friction itself are few and simple.

(Student) Friction is a force that opposes motion. Friction is a positive good when used in walking, stopping a car, and otherwise controlling motion, but is undesirable when it reduces the efficiency of machines. The amount of friction varies with the nature of the materials, the smoothness of the contact surfaces, and the force that holds the surfaces in contact. An attempt to relate a particular material to its property of producing friction resulted in the invention of the *coefficient of friction*. The coefficient of friction is

Friction

simply the ratio of the force necessary to overcome friction and the force holding the surfaces together called the *normal force:*

$$u = \frac{F_f}{N}$$

In Figure 12-1, F_f represents the force necessary to overcome friction and N represents the normal force. The normal force in the case of horizontal motion is simply the weight of the wood block plus any weight added to it.

PROCEDURE

Part A

Weigh a block of wood, lay it on a *clean* table with the largest side down, and after adding a 200 g. weight, draw the block across the table at a uniform rate of speed using a spring balance as shown in Figure 12-1. Record F_f as indicated by the spring balance reading.

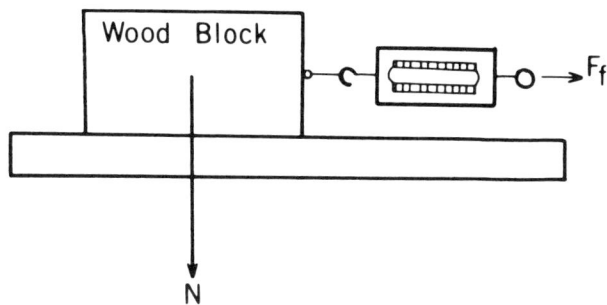

FIGURE 12-1

1. Repeat this five times and find the average F_f.
2. What is the coefficient of friction for this system?

Part B—The Effect of Surface Area upon Friction

Now rest the same block of wood as used in Part A on its smaller edge. Stop! Before going any farther, write an hypothesis which predicts the relationship of the force of friction and surface area.

Hypothesize that the force of friction will decrease, increase, or remain the same with the smaller surface area. Now add the 200 g. weight to it and draw it across the table with a spring balance. (Note: The general nature, or texture, of the surface of the smaller side must be the same as that of the larger side.) Record six trials and find the average force (F_f).
1. How does the average force of friction of Part B compare with the average force of friction of Part A?
2. Do you accept or reject your hypothesis concerning surface area and friction?
3. Determine the coefficient of friction from the data of Part B.

Part C—The Effect of the Normal Force upon Friction

Lay the wood block on its larger side, place 500 g. upon it. Stop! Before going any farther, write an hypothesis which predicts the relationship between the force of friction and the normal force. Now draw the block across the table at a uniform rate. Record six trials and find the average force.
1. How does the average F_f compare with the average F_f of Part A?
2. Do you accept or reject your hypothesis concerning normal force and friction?
3. Determine the coefficient of friction from the data of Part C.

Part D

Compare the coefficients of friction as determined under varying conditions in parts A, B, and C. Considering the experimental conditions, do you consider the three ratios as constant? (Teacher: A class discussion covering the interpretation of data is helpful here. The discussion should include such things as accuracy of measurement and constancy of experimental conditions. They may be asked whether the gross changes in surface area and normal force produced commensurate gross changes in the coefficient. It

The Inclined Plane

may be thus shown that generalizations may be drawn from comparatively rough data.)

DISCUSSION OF OUTCOMES

Part A
1. Student answer.
2. Student answer (u is a decimal usually in the range .3 to .5)

Part B
1. The forces of friction should be approximately the same.
2. Student answer.
3. Student answer. (u should be reasonably similar to the u determined in Part A).

Part C
1. The force of friction should be larger than in Part A.
2. Student answer.
3. Student answer. (u should be reasonably similar to the u determined in Part A).

Part D
The coefficient of friction is a constant for two surfaces. The coefficients derived from data in Parts A, B, and C should be similar.

EXERCISE 13—THE INCLINED PLANE

APPARATUS

1. Inclined Plane
2. Frictionless Car
3. Weights
4. String
5. Platform Balance
6. Meter Stick
7. Weight Pan

(Teacher) In Exercise 13 the emphasis is on the technological aspects of the inclined plane. The students will work with such concepts as: Actual Mechanical Advantage, Ideal Mechanical Advantage, and Efficiency. After working through these relationships once, the student is asked to hypothesize the effect of a second angle of incline upon these ratios, and to test his hypotheses. The skills reinforced with this exercise are hypothesizing by deduction, controlling and manipulating variables, experimenting, and interpreting data.

(Student) Raising heavy objects is made easier with the use of the inclined plane. The inclined plane is simply a sloped surface whereupon an object may be slid or rolled. Neglecting friction, the amount of force required depends upon the weight of the object and the slope of the incline. The force necessary to push the car in Figure 13-1 up the incline plane compared to the greater force necessary to raise it directly against gravity (its weight) is the *Actual Mechanical Advantage*:

$$\text{A.M.A.} = \frac{\text{(W) Weight of Object}}{\text{(F) Force necessary to push the object up the incline.}}$$

The *Ideal Mechanical Advantage* of an inclined plane is simply the ratio of the length of the incline to the height through which the object is raised:

$$\text{I.M.A.} = \frac{L}{H}$$

How well a machine does work without loss due to friction is called its efficiency and may be found with the following relationship:

$$\text{Efficiency} = \frac{\text{A.M.A.}}{\text{I.M.A.}}$$

Another method of determining efficiency is to compare work input to output:

$$\text{Efficiency} = \frac{\text{Work Output}}{\text{Work Input}} = \frac{\text{Wt. of Object} \times H}{\text{Force Up Incline} \times L}$$

The Inclined Plane

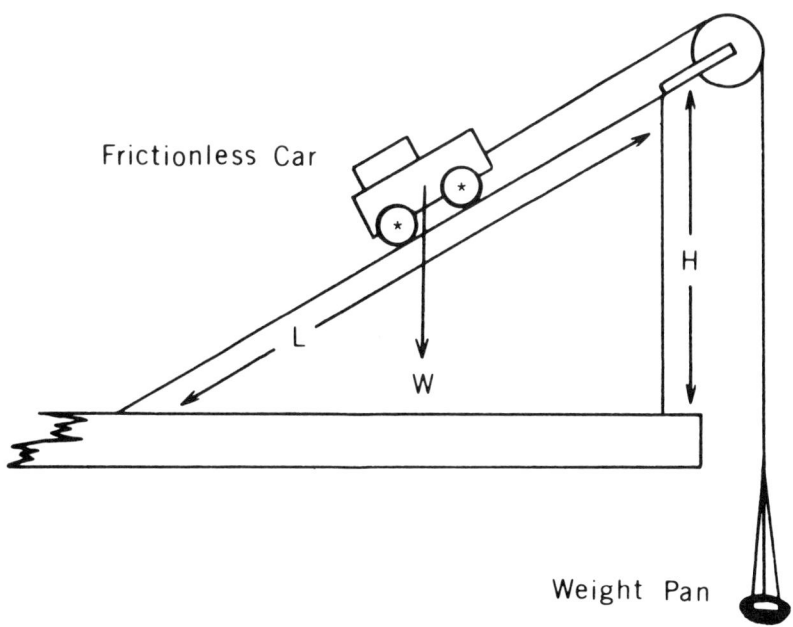

FIGURE 13-1

PROCEDURE

Part A—The Inclined Plane at 30°

After raising an inclined plane to an angle of 30° to the horizontal, place on it a frictionless car to which has been added a 1000 g. weight. A string attached to the car should be placed on a pulley and weight pan attached to the other end as Figure 13-1. As weights are added to the weight pan, firmly tap the car until it is pulled up the incline at a slow but uniform velocity. From this data determine:

1. The A.M.A. of this machine.
2. The I.M.A. of this machine.
3. The efficiency of this machine.

Part B

Write out hypotheses for each of the ratios (A.M.A., I.M.A., and Efficiency) for an incline of 45°. (Example: The A.M.A. of an inclined plane will be greater for an incline of 45° than for an incline of 30°.) Test these hypotheses. Write a description of the activities employed in testing your hypotheses and discuss any discrepancies that arise.

Part C

1. How much work was done in pushing the car up the 45° incline?
2. What is the magnitude of the increase in the potential energy when the car is at the top of the incline?
3. Compare the work done on the car with the increase in its potential energy.

DISCUSSION OF OUTCOMES

Part A

1. Student answer (Should approximate 2 to 1).
2. Determined by measurement. (Should approximate 2 to 1.)
3. Student answer and will vary with the amount of friction in the system.

Part B

The 45° incline should yield a smaller A.M.A. and I.M.A. The efficiency will remain fairly constant.

Part C

1. Student answer found by: Work = force × distance.
2. Student answer found by: P.E. = weight × change in height.

EXERCISE 14—HOOKE'S LAW

APPARATUS

1. Upright Support
2. Clamp
3. Meter Stick
4. Paper Pointer
5. Weight Pan
6. Weights
7. Coil Spring
8. Balance
9. Unknown Mass

(Teacher) Exercise 14 affords the opportunity to present to the students a concrete experience in the interrelationship of theoretical abstraction, observable phenomena, and technological adaptation. Following a discussion of the theoretical aspects of Hooke's Law, the student experiments to find that the stretch (strain) of a spring is proportional to the weight (stress) applied to it. He then is asked to use this relationship to determine the weight of an unknown object. In this exercise the student employs most of the primary processes as well as the following integrated processes: controlling and manipulating variables, experimenting, and interpreting data.

(Student) At any given temperature a body assumes a particular size because of the interaction of the attractive forces which hold the atoms of the substance together and the kinetic energy which causes the atoms to move apart. When no exterior force is applied, the interatomic distances are maintained so as to guarantee maximum stability of the material. When an exterior force called *stress* is applied, the ideal interatomic distances are shortened or lengthened producing an effect called *strain*. Up to the point where the distance between the atoms is permanently changed, called the *elastic limit*, the strain produced on a material is proportional to the stress applied.

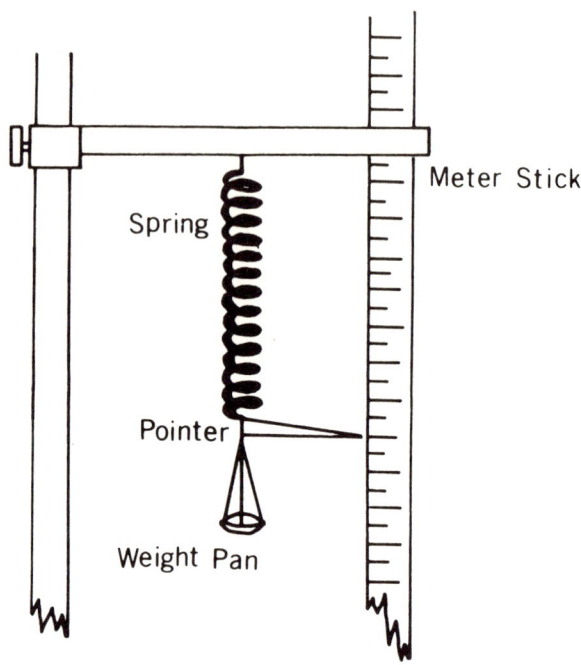

FIGURE 14-1

PROCEDURE

Part A

Clamp a meter stick to a horizontal bar in an upright position (see Figure 14-1). Suspend a metal spring with an attached weight pan to the horizontal bar. A piece of stiff paper or cardboard may be fashioned into a pointer and attached at the top of the weight pan. With no weights in the pan take a *no-load* reading from the meter stick where the pointer indicates. Add enough weight, say 50 grams, to stretch the spring a reasonable amount, and note the reading on the meter stick.

   1. A 50 g. stress produces how many cm. of strain upon the spring?

Hooke's Law

2. Remove the 50 g. weight. Did the pointer return to the no-load reading? What does this mean?
3. Complete the following chart. Check after each trial to see that the elastic limits of the spring were not exceeded.

Weight Added	Reading	Each Reading Minus No-Load Reading Stretch	Stretch Divided by Weight— Stretch per Gram of Wt.
0 g.	————	———————————	———————————
50 g.	————	———————————	———————————
100 g.	————	———————————	———————————
150 g.	————	———————————	———————————
200 g.	————	———————————	———————————
250 g.	————	———————————	———————————

Any graduated series of weights may be used if the above suggestions are not appropriate to a spring.

Put the above data in graphical form with "Weight Added" as the ordinate values and the "Reading" which indicates the values of the stretch of the spring as the abscissa.

4. What is the general shape of the curve?
5. Does the relationship that stress is proportional to strain remain true over the whole range of stress?

Part B

Remove the weight pan from the spring and attach an object of unknown mass. This mass must be such that the pointer indicates a reading somewhere between the no-load reading of Part A, and the place where the elastic limit is exceeded. Can you determine the weight in grams of the object? Check your finding against a reading on a platform balance.

DISCUSSION OF OUTCOMES

Part A

1. Student answer.
2. The pointer should return to the no-load reading. This indicates the elastic limit of the spring was not exceeded.

80 *Mechanics of Solids*

3. The *Reading* column and the *Stretch* column answers will vary with the characteristics of the spring. The answers in the *Stretch per Gram of Weight* column should be fairly constant.
4. The graphical representation should approximate a straight line.
5. The *Stress is proportional to strain* relationship should hold over the entire range of readings.

Part B

The student already has established a relationship which demonstrates the grams of stress required for each centimeter of strain in the spring. The weight of any object may be determined by multiplying the stretch of the spring in centimeters times the established grams per centimeter constant.

EXERCISE 15—THE WHEEL AND AXLE

APPARATUS
1. Wheel and Axle
2. Heavy String
3. Weights
4. Meter Stick
5. Weight Pan

(Teacher) The wheel and axle is a simple machine which is a component of many complex mechanical devices. Engineering concepts of mechanical advantage are easily illustrated with this device. In Part C of the exercise, the student is asked to design a device that yields maximum movement of a resistance. In this way he may see that force advantage is not always the desired characteristic of a machine. Technological considerations predominate in this exercise. However, the student experimenter is called upon to adapt certain relationships to a new use. This certainly calls for an exercise in insights valuable to problem-solving situations. In testing his insights, the student becomes an experimenter and gains experience in manipulating laboratory apparatus.

The Wheel and Axle

(Student) A wheel and axle is simply a large wheel fixed firmly to a small wheel so that a force applied so as to cause one wheel to rotate will cause the other to rotate also. If the wheel (W) has three times the diameter of the axle (A), one revolution of the system will cause the effort (E) to move three times as far as the resistance (R). Since work input equals output in any machine (neglecting friction losses), effort times effort-distance equals resistance times resistance-distance. If the effort in Figure 15-1 moves three times as far as the resistance, the amount of the effort required is only one-third of the resistance. The Ideal Mechanical Advantage (I.M.A.) of a wheel and axle is simply a ratio of the wheel's diameter to the axle's diameter. Because of friction losses, the Actual Mechanical Advantage (A.M.A.) is always somewhat less than the ideal. To find the Actual Mechanical Advantage, use the ratio of resistance to effort.

PROCEDURE

Part A

Set up a wheel and axle apparatus as shown in Figure 15-1. With a meter stick, find the diameter (or radius) of the wheel and of the axle of the apparatus.

What is the ideal mechanical advantage of this machine?

Part B

Wind a string about the *axle* and attach a 1000 g. weight for a resistance. Then wind another string in the *opposite direction* about the wheel and tie a weight pan to the end as shown in Figure 15-1. Now add weights to the weight pan and at the same time firmly tap the resistance in an upward direction until the resistance is raised at a uniform velocity. Find the Actual Mechanical Advantage of the machine. Compare the I.M.A. and A.M.A. to find the Efficiency of the machine.

Part C

A device is required that will produce maximum vertical movement of a resistance over a short distance. Draw a diagram of

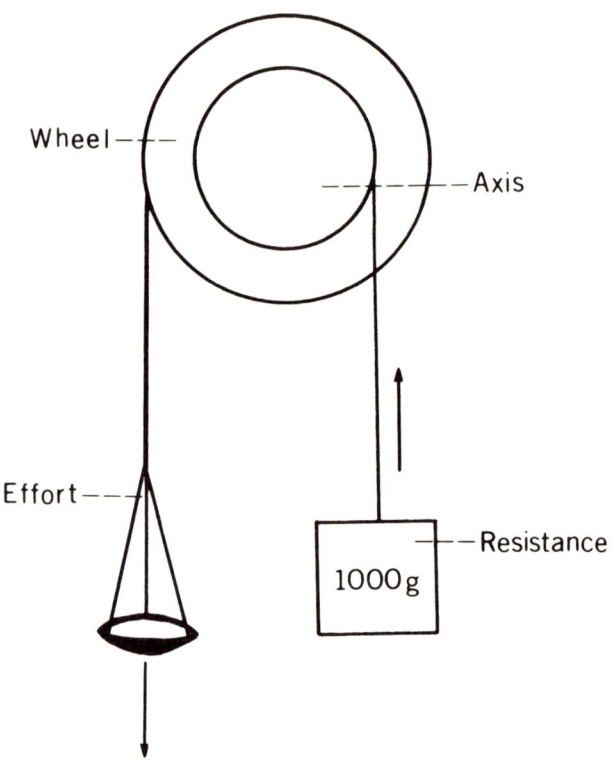

FIGURE 15-1

such a device and test your design by experimentation. Describe the reasoning that went into the design. How would you describe the mechanical advantage of the machine?

DISCUSSION OF OUTCOMES

Part A

$$\text{I.M.A.} = \frac{\text{Diameter of the wheel}}{\text{Diameter of the axle}}$$

Velocity and Acceleration 83

Part B

$$\text{A.M.A.} = \frac{\text{Resistance}}{\text{Effort}}; \quad \text{Efficiency} = \frac{\text{A.M.A.}}{\text{I.M.A.}}$$

Part C

A machine that will produce a maximum vertical movement may take the form of a wheel and axle with the resistance on the wheel or a lever with the fulcrum closer to the effort.

EXERCISE 16—VELOCITY AND ACCELERATION

APPARATUS
1. Buzzer Timer
2. Paper Tape
3. D.C. Source
4. Frictionless Car
5. Wood Block and Clamp
6. String
7. Weight Pan
8. Weights
9. Wheel in Clamp

(Teacher) Exercise 16 involves two of the most difficult manipulative tasks of a high school psysics laboratory. The first is to design an apparatus that will permit the investigation of the acceleration of a body. The second task is to illustrate for the student the power of graphical analysis as an experimental tool. Considerable manipulative skills are required for the student to successfully complete this exercise; thus he is strengthening this important aspect of dealing with science in the laboratory. The chief benefit to be derived, however, is that of learning to interpret data with the employment of a graphical tool. A new student concept of time may be a concommitant result of working with this exercise since velocity and acceleration are dealt with without resorting to the use of any conventional measure of time. Marks on a paper timer become the standard for this important dimension. Special paper tapes for this exercise may be purchased from most commercial suppliers. However, some buzzers will mark adding machine tapes without the need for carbon paper or a special waxed surface.

(Student) Longer intervals of time such as the year and day are readily determined as functions of the periodic movements of heavenly bodies. Water clocks, hour glasses, or mechanical timepieces serve to measure hours and minutes of time. Seconds and tenths of seconds may be determined with some accuracy using a stop watch. For laboratory purposes, an electric bell buzzer may be adapted to measuring time intervals in the range of hundredths of seconds. It is required in this exercise to adapt the strikes of a buzzer upon a paper tape to the study of changes in velocity of bodies with the application of constant forces.

PROCEDURE

Part A

Set up the apparatus as shown in Figure 16-1, with the paper tape inserted through the timer. As the car is moved forward, the tape will be drawn through the timer and have imprinted upon it a mark for each oscillation. The mark should occur at equal *time* intervals. Place enough weight in the weight pan to cause a smooth acceleration of the car, and with the timer operating, release the car.

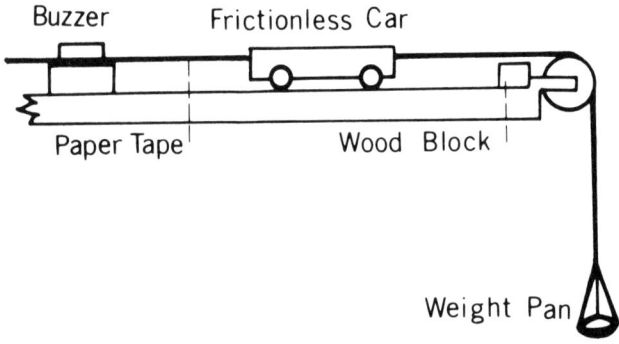

FIGURE 16-1

Velocity and Acceleration

1. What supplies the constant force used to accelerate the car?

Choosing any mark as zero, number the next five marks on the tape consecutively and measure the distance between them. Plot this information on an axis, as in Figure 16-2, and draw a smooth line that best describes the pattern of the plots. At the 2 mark point draw a line tangent to the curve and construct the horizontal and vertical lines t and D. The velocity of the car at this point is equal to the ratio of the distance per unit of time, $\frac{D}{t}$. D is in cm. and t in marks.

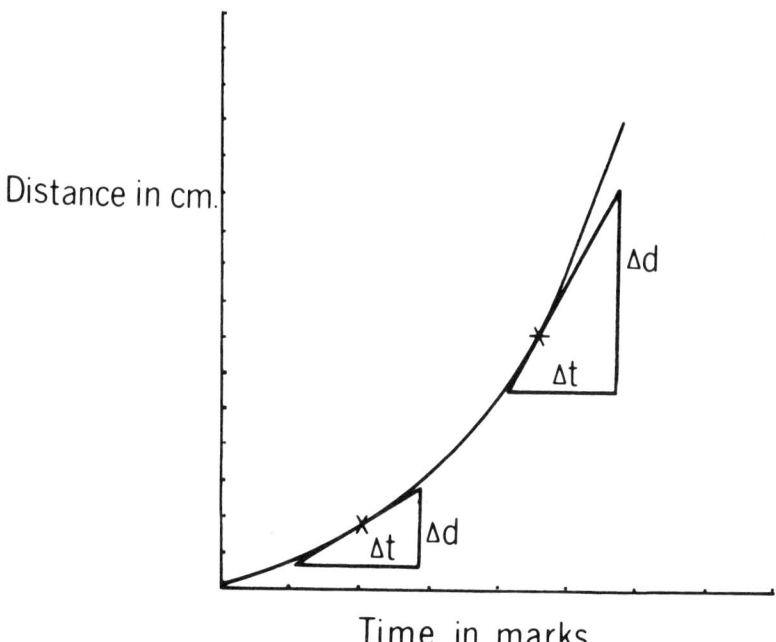

FIGURE 16-2

2. What is the velocity at the two strike point?
   (Remember the unit of velocity in this instance is cm./mark).

3. Find the velocity at the five strike point on the curve in the same way.
4. How much did the velocity increase in the 3 mark time interval?
5. What is the acceleration rate of the car in cm./mark²?

Part B

Repeat this exercise using a different accelerating force.

DISCUSSION OF OUTCOMES

Part A

1. The accelerating force is supplied by the gravitational attraction of the earth for the weight pan and weights.
2. Student answer.
3. Student answer.
4. Student answer found by subtracting answer 2 from answer 3.
5. Student answer found by dividing the increase in velocity between points 2 and 5 by a factor of 3.

Part B

All operations for Part B are similar to those of Part A.

EXERCISE 17—CONSERVATION OF MOMENTUM

APPARATUS

1. Timer
2. Paper Tape
3. Frictionless Car
4. Two Bricks
5. Wood Block and Clamp
6. Upright Stand

(Teacher) Exercise 17 deals with a fundamental physical principle, the Conservation of Momentum. As in most exercises

Conservation of Momentum

requiring a measure for velocity, success is dependent in large measure upon the manipulative skill of the experimenter. Patience and careful attention to detail are necessary here, virtues that are possessed by most competent laboratory technicians. As in Exercise 16, the student will be required to read and interpret data from a timer-marked tape. In general the skills to be strengthened in this exercise are those of experimenting and interpreting data.

(Student) Momentum is said to be the amount of motion of an object, and is a constant where no unbalanced forces act upon the object. The momentum is defined by the product of the mass and the velocity (MV). This experiment attempts the study of the Law of Conservation of Momentum by adding mass to a moving body and observing the effect upon its velocity. If the law holds true, one should observe the adjustment of the velocity of a freely moving body so that the momentum remains a constant.

PROCEDURE

Part A

Suspend a brick with string so a cart can barely pass under it (see Figure 17-1). The brick must be horizontal and motionless. With the timer activated, give the car a push to give it a reasonably high velocity, and as it passes under the brick, cut the string to allow the brick to fall upon the car.

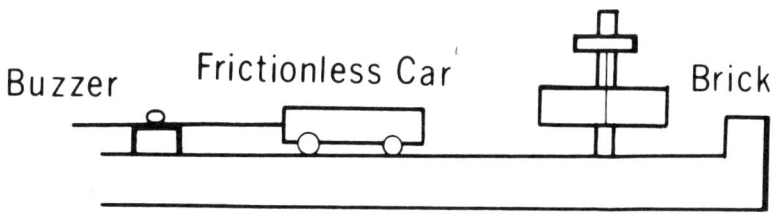

FIGURE 17-1

1. From the tape, can you find the average velocity in cm./mark before the brick was added?

2. Can you find the average velocity in cm./mark after the mass of the brick was added?
3. After weighing the car and brick find the momentum before and after the addition of the brick's mass.
4. Does this substantiate the Law of Conservation of Momentum?

Part B

With one brick in the cart at the start, repeat the procedures of Part A.
1. Does this second trial involving different data reinforce the findings of Part A?

DISCUSSION OF OUTCOMES

Part A

1. Student answer as determined by finding the average number of centimeters the car travels in a time interval over a several mark period on the paper tape.
2. Student answer found by the procedure of item 1.
3. Student answer determined by multiplying the masses before and after the addition of the brick by their respective velocities.
4. The momentum before and after the addition of the brick should be approximately the same.

Part B

1. A repeat of the procedures of Part A with different masses should yield the same general results.

EXERCISE 18
OPTIONAL PROBLEMS TO TEST LABORATORY SKILLS

(Teacher) This exercise represents the climactic ending for Unit II. It is a test of the ability of the students to apply the process

Optional Problems 89

skills, manipulative skills, and subject matter content in an independent attack upon a problem. The test should be issued as a challenge with a proper amount of exhortative introductory remarks.

These problems are presented to the student along with suggestions on how to begin the work which may lead to their solutions. The problems lend themselves to investigations which call for skills which have been developed in previous exercises, with apparatus already familiar to the student. It is especially important for the teacher to remind himself of his goals during this phase of the laboratory program. His primary concern is the development by his students of the skills, processes, and attitudes of the independent researcher. To best realize his objective the teacher must provide the leadership to help the students to maintain some degree of momentum in their investigations, but he must refrain from supplying direct answers to their questions. Remember—you are interested in processes not data.

(Student) It is time for you to attempt the solution of a laboratory problem on your own. Three suggestions for problems to be solved follow. In general your investigation may be conducted by employing the apparatus and materials available in the laboratory. However, feel free to employ any additional equipment which you may feel will aid in your work.

Let us see how well you can operate as an independent investigator.

PROCEDURE

Problem A—Conservation of Momentum—Collision

Proposal: Permit a swinging heavy pendulum bob to collide with a hanging lighter pendulum bob. Velocity data may be obtained by attaching paper tapes to the bobs and running them through a timer of the buzzer type.

An alternate method may be to show that the velocity of a pendulum bob at the equilibrium position is proportional to the square root of the height through which it was raised. The mathematics follows:

The kinetic energy ($\frac{1}{2}$ mv^2) of the bob at the equilibrium position equals the potential energy (mgh) at the highest position:

$$\frac{1}{2} MV^2 = Mgh; \quad V = \sqrt{2gh}$$

The Conservation of Momentum states $M_1 \cdot V_1 = M_2 \cdot V_2$. There is no loss of momentum because of collision. So:

$$M_1 \sqrt{2gh_1} = M_2 \sqrt{2gh_2}$$

$$M_1 \sqrt{h_1} = M_2 \sqrt{h_2}$$

Can you prove the Law of Conservation of Momentum experimentally?

Problem B—Newton's Second Law of Motion

Is the acceleration of a body proportional to the force applied upon it? Is the acceleration inversely proportional to the mass of the body? Using a weight pan hung over the table's edge to supply a constant force, the time (t) it takes for a frictionless car of mass (m) to proceed from rest across distance (s) may be determined. If $S = \frac{1}{2} at^2$, then $A = \dfrac{2S}{T^2}$. Can you use this in the proof of Newton's Second Law?

An alternate method is to use a timer and paper tape to acquire velocity, acceleration data.

Problem C—The Relationship of Potential Energy, Kinetic Energy and Work

One method of investigating the relationship of potential energy (mgh), Kinetic energy ($\frac{1}{2} MV^2$), and work (F × S) is to place a car of known mass upon an incline at height h, above the table. Let the car roll down gaining kinetic energy until at the bottom of the incline it is caused to drag wood blocks a distance (S) before it stops. The force of friction exerted upon the sliding blocks may be found (F = uN) after u has been determined. Does the work done in stopping the car equal the potential energy of the car on the incline? How is the kinetic energy related to the potential energy and the work done? Can you find the velocity

Optional Problems

of the car at the bottom of the incline? Remember, the procedures outlined above are suggestions only.

Use any method that seems likely to offer the possibility of achieving the desired goal. Can you accept this challenge?

DISCUSSION OF OUTCOMES

Problem A

When students use the collision of the pendulum bobs method for testing the Conservation of Momentum they must account for the total momentum (energy) after the collision. This means they must determine the maximum height of each bob after they collide. Results with this method of dealing with Problem A tend to be rather good.

Problem B

The secret to achieving success with Problem B appears to reside in the selection of the proper amount of weight so as to cause the frictionless car to accelerate slowly but evenly. The students may be encouraged to time many trials with a stop watch that they may be assured of the accuracy of the time factor.

Problem C

Problem C generally provides the greatest difficulty of the three suggested problems. The results appear to be affected appreciably by the starting friction of the braking blocks. The challenge for the student is to suggest the reason for the discrepancy of his results.

Unit III

Mechanics
of
Liquids and Gases

Pressure of Liquids
The Law of Buoyancy
Density and Specific Gravity
Specific Gravity of Liquids
Atmospheric Pressure
Optional Problems to Test
  Laboratory Skills

Unit III

Some teachers may wish to think of the exercises of Unit III as supplements to Unit II. The activities of the unit deal with concepts common to many high school textbooks dealing with the topics of liquids and gases. They provide the student with additional practice in using the metric system and illustrate several procedures for producing indirect measurement. The teacher may wish to select several of the exercises to be used as classroom demonstrations and eliminate Unit III as a block of laboratory exercises.

EXERCISE 19—PRESSURE OF LIQUIDS

APPARATUS
1. Large Test Tube
2. Lead or Copper Shot
3. Metric Rule
4. Set of Metric Weights
5. Saturated Salt Solution

(Teacher) Exercise 19 provides the student further opportunity to work with metric units. The activity also develops an understanding of the concept of liquid pressure, and contributes to the differentiation of the two often confused terms, force and pressure. The process skills required of the student in performing the exercise would appear to be those of experimenting and interpreting data.

(Student) That pressure of a liquid increases with depth may be illustrated with the following example. The force that supports a boat in the water is exerted by the water upon the bottom of the boat. The total force necessary to support the boat is equal to the weight of the boat. The pressure on any unit area of the boat bottom is equal to the force on the bottom divided by the area of the bottom. If weight is added to the boat it sinks deeper into the water until the buoyant force exerted on the bottom increases sufficiently to regain equilibrium. Since the area of the bottom is a constant, the value of the pressure, $\dfrac{\text{Force}}{\text{Area}}$, is the factor that becomes greater.

This line of reasoning may be used to prove the pressure exerted by a liquid is directly proportional to the depth of the liquid.

PROCEDURE

Part A

Place enough lead or copper shot in a test tube so that it floats upright in tap water. (Melted paraffin will hold the shot in place if necessary.)

1. Measure the distance from the surface of the liquid to the bottom of the test tube. Add a small weight by sliding it *gently* down the side of the tilted test tube.
2. What is the depth now?
3. How much did it sink per gram of added weight? Add two more weights to the test tube by the procedure described above.
4. What is the depth/gram increase of all three weight increases?
5. Since the horizontal area upon which upward forces act remains the same, how do you account for the fact the tube continues to float?

Part B

Repeat Part A using a saturated salt water solution.
1. What is the increase in depth per gram of weight?
2. Is the average depth increase per gram as great for salt water as for fresh water?
3. Why?
4. The pressure exerted by a liquid is proportional to what two factors?

DISCUSSION OF OUTCOMES

Part A

1. Student data.
2. Student data.
3. Student data.
4. Student data.
5. The force exerted to support the increased weight is greater, therefore the upward force exerted on each unit area of the test tube is greater.

Part B

1. Student data.
2. No. The average depth increase is less per gram of weight.
3. The pressure exerted by a liquid is proportional to both the depth and density of the liquid.

EXERCISE 20—THE LAW OF BUOYANCY

APPARATUS

1. Platform or beam balance
2. Block of heavy metal
3. Block of Wood
4. String
5. Overflow Can
6. Catch Bucket
7. Salt water

(Teacher) The Law of Buoyancy has many applications to the experiences of most high school students. Young people readily recall the feeling of floating while swimming, of supporting heavy objects while they are immersed in water, and of watching a boat sink lower and lower as more and more passengers climb aboard. Exercise 19 is designed to bring understanding of some of the physical relationships involved in their observations. To complete the activity the students will be asked to experiment, interpret data, and to generalize upon their data.

(Student) Why may swimmers float in water? Why may heavy rocks be raised to the surface of a lake and then become impossible to raise further? You may guess the answer to these two questions has something to do with a phenomenon called buoyancy. This exercise is designed to help you to answer such queries and to increase your understanding of the Law of Buoyancy.

PROCEDURE

Part A

Fill an overflow can up to the spout with water and then carefully add more water until it is just ready to pour from the spout.
1. Find the weight of a dry overflow can. Now suspend a block of heavy metal from a platform or beam balance by means of a string.
2. What is the weight of the metal block in air? Now completely immerse the metal block in water while it is still suspended from the balance. Allow the excess water to flow into the catch bucket as in Figure 20-1.
3. Find the loss of weight of the object.
4. What is the weight of the displaced water (weight of catch bucket plus water minus the weight of the bucket)?
5. How does the loss of weight of the immersed solid compare with the weight of the displaced water?

Part B

Repeat Part A using salt water.
1. How does the loss of weight of the object compare with the weight of the displaced salt water?
2. How does the loss of weight in salt water compare with the loss in fresh water?
3. Explain.

Part C

Fill an overflow can with fresh water until it is just ready to flow out of the spout. After placing a catch bucket under the over-

The Law of Buoyancy

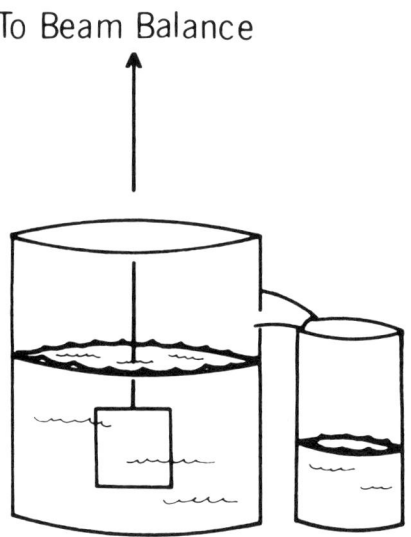

To Beam Balance

FIGURE 20-1

flow spout place an object that will float in the overflow can. (A piece of light wood will do.) Weigh the displaced water and the floating object.

1. How do their weights compare?
2. Can you make a generalization regarding the buoyant force exerted by a liquid upon an object and the weight of the displaced liquid?
3. What determines whether a body will float or sink in a liquid?

DISCUSSION OF OUTCOMES

Part A

1. Student data.
2. Student data.
3. Student data.
4. Student data.

5. The loss of weight of the object when immersed in the water should approximate the weight of the displaced water.

Part B

1. The loss of weight of the object when immersed in the salt water should approximate the weight of the displaced salt water.
2. The loss of weight in salt water is greater (perhaps by a factor of .1).
3. The same amount of liquid is displaced. However a volume of salt water weighs more than an equal volume of fresh water.

Part C

1. The weight of the displaced water should be approximately the same as the weight of the floating object.
2. The generalization is that the buoyant force exerted upon an object is equal to the weight of the displaced liquid.
3. A body that weighs more than an equal volume of water will sink; a body that weighs less than an equal volume of water will float.

EXERCISE 21—DENSITY AND SPECIFIC GRAVITY

APPARATUS

1. Platform or beam balance
2. Metric ruler
3. Spring
4. Metal Object
5. Light Object (paraffin or wood block)
6. Beaker
7. Overflow Can
8. Catch Bucket

(Teacher) The concepts of density and specific gravity as definitions of properties of substances have long been troublesome to

Density and Specific Gravity

the beginning physics student. Exercise 21 is designed to bring about understanding of these two very common measures through the manipulation of materials. In this way the nature and interrelationships of density and specific gravity as useful laboratory and industrial tools may become meaningful to the student. The student will be required to define terms, formulate deductive hypotheses, experiment, and interpret data.

PROCEDURE

Part A

1. Carefully measure a metal cube or rectangular solid using a metric ruler, and find its volume in cm.^{3}.
2. Now find the mass in grams of the solid on a platform or beam balance.
3. Can you estimate what one cubic centimeter of the metal substance would weigh?

The mass (in grams) per cubic centimeter defines the *density* of the metal block. Then:

$$\text{density of a solid} = \frac{\text{mass of the solid}}{\text{volume of the solid}}$$

4. What would be a proper unit for density in the English system of measurement?

The relationship *mass per unit volume* is also used to describe the density of liquids with volumetric measurements being made directly with the use of graduated vessels.

Specific gravity is a ratio used to compare the density of a substance to the density of water, and so may be represented by the expression $\frac{\text{Density of a substance}}{\text{Density of water}}$. In experimental situations the following expressions may be used to determine the specific gravity:

$$\text{Sp. Gr.} = \frac{\text{Weight of Substance}}{\text{Weight of Equal Vol. of Water}}$$

$$\text{or } \frac{\text{Wt. of Solid}}{\text{Loss of Weight in Water}}$$

Specific gravity constants are used for identification of unknown substances and for determining their buoyant properties.

Part B—Specific Gravity of Heavy Solids

1. Suspend a piece of metal from a beam or platform balance and determine its weight.
2. Find the loss of weight when the object is immersed in water.
3. Using the expression found in the introduction of this experiment, which involves the weight of an object and its loss of weight in water, find the specific gravity of the piece of metal.

Part C—Specific Gravity of Heavy Solids, Alternate Method

Fill an overflow can to the spout so that the addition of a drop of water will cause a drop to flow from the spout. Find the weight of a dry catch bucket and place it under the spout of an overflow can. Immerse the metal object used in Part B in the overflow can.

1. What is the weight of the displaced water?
2. Using the relationship found in the introduction which involves the weight of an object and the weight of an equal volume of water find the specific gravity of the metal object.
3. How does this compare with that of Part A?
4. What does the number that represents the specific gravity of the metal block mean?

Part D—Specific Gravity of Light Solids

To determine the specific gravity of a light solid, a heavy sinker must be attached to the light solid and both of the objects suspended from a balance so that the sinker is immersed in the water of an overflow can while the light solid is in the air (see Figure 21-1). Fill the overflow can to the spout, place a catch bucket of

Density and Specific Gravity

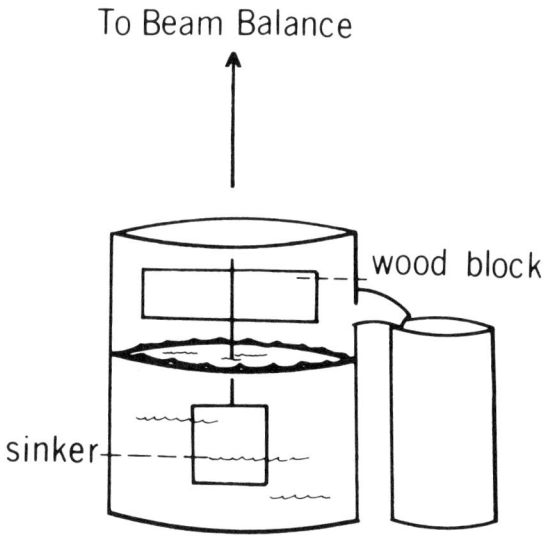

FIGURE 21-1

known weight under the spout, and allow the light object to be completely immersed in the water.

1. Find the weight of the displaced water and the light solid.
2. Using the relationship which involves the weight of an object and the weight of an equal volume of water find the specific gravity of the light solid.
3. What is the specific gravity of water?
4. How may the specific gravity of an object allow one to predict the result of placing the object in a volume of water?

DISCUSSION OF OUTCOMES

Part A

1. Student data.
2. Student data.

3. Student answer found by reducing the expression:

$$\frac{\text{weight of the object in grams}}{\text{volume of the object in cm.}^3}$$

4. Common English density units are lbs. per ft.³, or oz. per in.³

Part B

1. Student data.
2. Student data.
3. Student answer found by reducing the expression:

$$\frac{\text{wt. of a solid}}{\text{loss of wt. in water}}$$

Part C

1. Student data.
2. Student answer found by reducing the expression:

$$\frac{\text{wt. of a solid}}{\text{weight of an equal volume of water}}$$

3. The specific gravity quotients as determined in Parts B and C should compare favorably.
4. The specific gravity coefficient relates the density of a substance as compared to the density of water.

Part D

1. Student data.
2. Student answer found by reducing the expression:

$$\frac{\text{wt. of a substance}}{\text{weight of an equal volume of water}}$$

3. The specific gravity of water is 1.00.
4. A substance with a specific gravity which is less than 1.00 will float in water while a substance with a specific gravity greater than 1.00 will sink.

Specific Gravity of Liquids

EXERCISE 22—SPECIFIC GRAVITY OF LIQUIDS

APPARATUS
1. Alcohol
2. Salt Water
3. Hydrometers (light and heavy liquids)
4. Beam Balance
5. Metal Sinker
6. Beaker
7. String
8. Hydrometer Jar

(Teacher) Exercise 21 is designed to provide the student with experiences that promote understanding of liquid density and specific gravity concepts. The exercise also demonstrates laboratory techniques for determining properties of substances. The process skills required of the exercise would appear to be those of experimenting, interpreting data, and understanding number relationships.

(Student) The specific gravity of a liquid is a coefficient that expresses the relationship: $\dfrac{\text{density of a liquid}}{\text{density of water}}$.

The specific gravity of a liquid may be determined by the ratio:

$\dfrac{\text{weight of a liquid}}{\text{wt. of an equal volume of water}}$ or by $\dfrac{\text{loss of weight of an object in a liquid}}{\text{loss of weight of the same object in water}}$

or by employing a hydrometer. The height of the floating hydrometer is determined by the density of the liquid in which it is placed.

PROCEDURE

Part A—Specific Gravity of Liquids by Equal Volumes Method

Weigh a dry specific gravity bottle (Figure 22-1), then fill it with water. Replace the stopper removing the excess water that escapes. Dry the outside of the bottle, weigh it and its contents, and record the data.

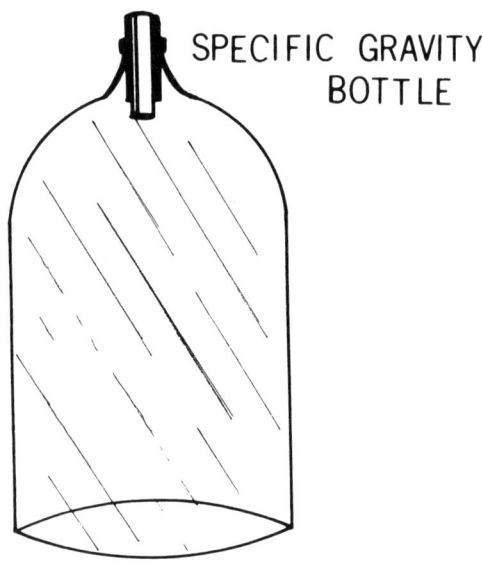

FIGURE 22-1

1. What is the weight of the water alone?
2. Now in the same way find the weights of the equal volumes of alcohol and salt water that fill the bottle.
3. Using the relationship from the introduction which involves the weight of a liquid and weight of an equal volume of water, find the specific gravity of alcohol.
4. Find the specific gravity of salt water.

Part B

Some hydrometers are made to use in liquids heavier than water, some in liquids lighter than water, and some in both.
1. Using the appropriate hydrometer determine the specific gravity of alcohol.
2. Salt water.
3. Tap water.

Specific Gravity of Liquids

Part C

Weigh a heavy solid and then suspend it from a platform or beam balance as it is immersed in water first, then alcohol, then salt water.
1. What is the loss of weight in water?
2. In alcohol?
3. In salt water?
4. Using the relationship found in the introduction involving loss of weight of an object in a liquid and the loss of weight of an object in water find the specific gravity of alcohol.
5. Salt water?
6. What is the percent of difference of the specific gravities of alcohol as found by the bottle and displacement methods?
7. By the bottle and hydrometer methods?
8. Can you find the specific gravity of mercury?

DISCUSSION OF OUTCOMES

Part A

1. Student data found by subtracting the weight of the dry bottle from the weight of the bottle filled with water.
2. Student data.
3. Student answer. $\dfrac{\text{wt. of alcohol}}{\text{wt. of equal volume of water}}$
4. Student answer. $\dfrac{\text{wt. of salt water}}{\text{wt. of equal volume of water}}$

Part B

1. Student data.
2. Student data.
3. Student data.

Part C

1. Student data.
2. Student data.
3. Student data.
4. Student answer. $\dfrac{\text{loss of weight of the object in alcohol}}{\text{loss of weight of the object in water}}$
   (approx. .9)
5. Student answer. $\dfrac{\text{loss of wt. of the object in salt water}}{\text{loss of wt. of the object in water}}$
   (approx. 1.1)
6. Student answer. $\dfrac{\text{difference between the two sp. gr. coefficients}}{\text{sp. gr. coefficient by the btl. method}}$
7. Student answer. $\dfrac{\text{difference between the two sp. gr. coefficients}}{\text{sp. gr. coefficient by the btl. method}}$
8. Student answer as determined by his own process. The most likely procedure is the bottle method.

EXERCISE 23—ATMOSPHERIC PRESSURE

APPARATUS
1. Vacuum Pump
2. Bell Jar
3. Bell Jar Platform
4. Balloon
5. Two Collection Bottles
6. One Hole Stopper
7. Rubber or Glass Tubing
8. Platform or Beam Balance
9. Weights
10. Magdeburg Hemispheres

(Teacher) Exercises dealing with atmospheric pressure are often performed as teacher demonstrations. Expensive equipment

Atmospheric Pressure

such as bell jars and vacuum pumps generally preclude the possibility of supplying each lab station with individual apparatus. However, with the system of blocking laboratory exercises, as is advocated by the author, it is possible to arrange to have several groups per day performing a given activity, the remainder of the class being involved with other exercises of the block.

Exercise 23 provides the student with experiences that make real the effects of atmospheric pressure. The students are required to predict the effects of reduced atmospheric pressure upon various systems, and to compute the somewhat startling magnitude of the total force which the atmosphere exerts upon a small sphere. Process skills required for the exercise would appear to be observing, predicting, inferring and deductive problem solving.

(Student) The earth's atmosphere is composed of gases. The accumulative effect of individual gas molecules striking a surface is exhibited as pressure upon the surface. The atmosphere exerts a force upon every surface with which it is in contact. Atmospheric pressure may be caused to do work in devices such as pumps and siphons. Normal atmospheric pressure will support 76 cm. of mercury or 32 ft. of water.

PROCEDURE

Part A

Inflate a balloon to only a fraction of its maximum size.
1. What are the forces acting upon the balloon that determine the size it assumes?
2. Predict what would happen if the balloon were placed in a vacuum. Place the balloon in a bell jar and evacuate the air from the jar.
3. Was your prediction correct?
4. Can you explain why the events occurred as they did?

Part B

Fill a collection bottle about two-thirds full of water and insert a one-hole stopper in the mouth of the bottle (see Figure 23-1). Insert glass tubing so that it extends nearly to the bottom of the

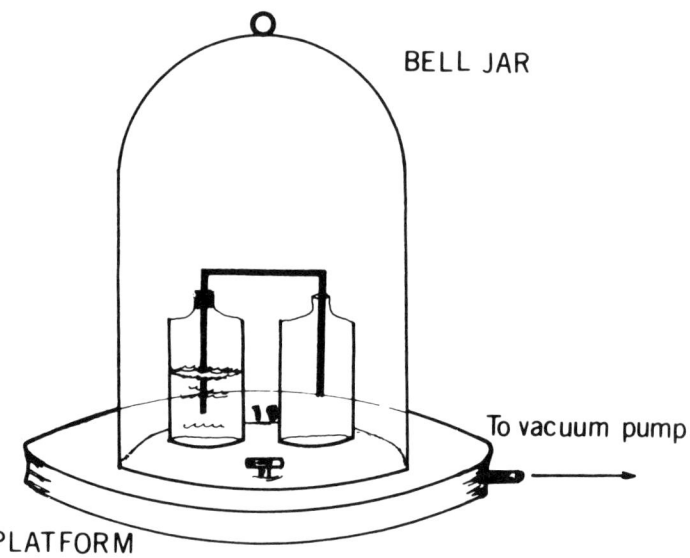

FIGURE 23-1

bottle. Use either rubber or glass tubing to extend nearly to the bottom of a second bottle which is open-mouthed.

1. Predict what would happen if the apparatus were placed in a vacuum, then place it in a bell jar and evacuate the air.
2. Was your prediction correct?
3. Explain the sequence of events.
4. Predict what will occur when the air is permitted to return to the bell jar, then permit the air to return.
5. Is your prediction correct?
6. Explain the sequence of events as the air returns to the bell jar.
7. How might this principle be applied to a system to raise water from a 30 ft. well?

Part C

Magdeburg hemispheres are halves of a sphere fitted with a stopcock through which the air may be evacuated when the halves

Atmospheric Pressure *111*

are placed together. Place the hemispheres together and weigh them carefully.
1. What is the weight of the hemispheres with the cavity filled with air? Evacuate the air from the sphere, close the stopcock, and on the same balance previously used weigh the sphere.
2. How great is the difference in the weight of the sphere?
3. The loss of weight may be attributed to what factor?
4. Try to pull the hemispheres apart. Result?
5. Assuming the atmospheric pressure inside the spheres to be zero (which of course, it is not) and the atmospheric pressure to be 14.7 lb./in.^{2} can you estimate the total force which is holding the sphere together?

DISCUSSION OF OUTCOMES

Part A

1. Forces acting upon the balloon are internal and external air pressures, and the elastic force of the balloon.
2. Student statement. The student should predict the expansion of the balloon.
3. Student answer. Yes or No.
4. The balloon expanded due to the contrived imbalance in the forces acting upon the balloon. The reduced external gas pressure permitted the expansion of the balloon until such time that the internal pressure, the external pressure and the elastic strength of the balloon again achieved equilibrium.

Part B

1. The student should predict the transfer of the liquid from the closed bottle to the open bottle and might also predict the loss of the gases from the closed bottle.
2. Student answer. Yes or No.
3. The reduced air pressure in the open bottle resulted in an imbalance in the system. The greater gas pressure in the

closed bottle forced the liquid from the closed bottle into the open bottle.
4. The student should predict the return of the liquid to the closed bottle.
5. Student answer. Yes or No.
6. The increased pressure upon the liquid in the open bottle forced the liquid from the open bottle into the closed bottle.
7. Atmospheric pressure will raise water through 30 plus feet in a system where a vacuum has been created.

Part C

1. Student data.
2. Student data.
3. The loss of weight is due to the loss of air in the sphere.
4. The hemispheres cannot be pulled apart.
5. The force exerted upon the outside of the sphere may be estimated by $F = 14.7/\text{lb.}/\text{in.}^2 \times 4\pi r^2$, where r is the radius of the sphere in inches ($4\pi r^2$ represents the area of the sphere).

EXERCISE 24
OPTIONAL PROBLEMS TO TEST LABORATORY SKILLS

(Teacher) The following problems may be assigned to the students as an exercise which allows them to develop independent attacks for the investigation of a problem in the laboratory. As in the previous exercises of this nature it is important that the teacher permit the students a maximum amount of freedom as they pursue their investigations. The products of the laboratory work during this exercise do not have near the importance as do the processes that are employed by the individual teams of investigators.

(Student) It is again time for you to pursue an independent investigation of a problem of science. In the instances where the principles or laws of science are known to you, you are to proceed as though you are challenging the laws with your investigation. Your teacher will announce to you the amount of infor-

Optional Problems *113*

mation you may have at your disposal to begin the investigation, and will suggest materials which you might use.

Accept the challenge this exercise offers you. Prove that you are capable of designing and carrying out your plan for investigating a problem in science.

PROCEDURE

Problem A—The Effect of Temperature upon the Volume of a Gas

Can you establish with an investigation of your own design the effect of temperature upon the volume of a gas? You should attempt to provide some sort of quantitative relationship, perhaps in the form of ratios of volumes and temperatures, or perhaps graphical data to show the relationship.

Problem B—The Effect of Depth of a Liquid upon Pressure

Can you establish experimentally the effect of depth of a liquid upon the pressure exerted by the liquid? Many people possess the intuitive belief that the pressure exerted by a liquid increases with the depth of the liquid. Your job is to provide experimental evidence to support or refute this belief. You should provide convincing data along with a description of your procedures in order to establish the likelihood of your position.

DISCUSSION OF OUTCOMES

Problem A

The students may do a creditable job of establishing the principle that gases expand upon heating with the use of a plastic bag and some sort of heater which does not employ a direct flame. The procedure may go something like this: Put a *small* amount of air into a plastic bag, seal it, and roll it tightly from one end trapping the air in the other end of the bag. A thermometer may be inserted into the neck of the bag if care is taken to seal around the stem with clay, gum, or some other substance. The volume of the gas may be estimated by measurement. Next heat the bag. Take temperature and volume data at regular intervals. You may refer the students to the Absolute temperature scale if you wish.

Some students may choose to use a balloon as their expandable container. The force necessary to do work on the balloon will introduce some error into the investigation. Do not instruct the student against the use of balloons, but be ready to suggest to them ways of eliminating this type of error, such as using balloons of large capacity and starting with a small volume of air.

Problem B

The simplest procedure for gathering evidence regarding the relationship of the depth of a liquid to the pressure exerted by the liquid is to put a single hole in the side of a container about six inches from its bottom and fill the container with water to various heights above the hole. Releasing a finger from the hole with the water at some known height above the hole will permit the water to be expelled from the can. The first water particles to emerge from the opening will follow a trajectory, the nature of which is determined primarily by the force exerted upon them as they leave the container. A plot of the distances from the can at which the water strikes a horizonal plane against the heights of the liquid above the opening should reveal a pattern that allows analysis of the relationship that exists between the two variables.

The process of relating the horizontal range of the water particles to the water pressure may go something like this:

1. Since the height of the opening above the horizontal plane is a constant the time of fall for the water particles for each experimental situation is identical.
2. The difference in the horizontal distances of the water particles may be attributed to differences in their average velocities.
3. The average velocity of each particle is dependent upon the acceleration of the particle, $V_{ave.} = \dfrac{V_f + V_1}{2}$, where $V_1 = 0$ and $V_f = at$, $V_{ave.} = \dfrac{at}{2}$

Optional Problems

4. Also the acceleration of a particle is directly related to the force exerted upon it, $F \propto a$.

5. Since pressure $= \dfrac{\text{force}}{\text{unit area}}$, the horizontal distance which a water particle ejected from an opening at some height above a horizontal plane travels is directly related to the pressure of the liquid at the point of the opening.

Some students may attack the problem of the relationship of depth of a liquid to its pressure by attempting to submerge balloons to different depths and somehow estimating the volume of the balloon at each depth. They should be cautioned that the pressure inside the balloon is not a constant, that it increases with each decrease in the size of the balloon. Often this consideration necessitates a somewhat more complex interpretation of their experimental data than the students initially surmised.

Unit IV

Wave Phenomena

Sound and Light

Frequency of a Tuning Fork
Determination of the Wavelength and
    Velocity of Sound by Resonance
Vibrations of Strings
Intensity of a Light Source
Reflections from a Plane Surface
Images in a Curved Mirror
Refraction of Light
Refraction in a Convex Lens
The Visible Spectrum
Waves in a Ripple Tank
Optional Problems to Test
    Laboratory Skills

Unit IV

EXERCISE 25—FREQUENCY OF A TUNING FORK

APPARATUS
1. Vibrograph
2. Watch or Timer
3. Candle or Camphor and Evaporating Dish

(Teacher) A laboratory exercise which employs the vibrograph to determine the frequency of a tuning fork provides the high school student experiences in several areas. The exercise serves to illustrate the commonly used procedure of producing tracings which serve as descriptive evidence of motion. The seismograph, tachometer recorders, and industrial recorders are examples of devices which utilize tracings to provide evidence for subsequent interpretation. It also shows one way in which an investigator may determine the characteristics of a motion which may not be examined by direct visual means. The method is to determine the relationship of the motion to a second motion which has properties more amenable to procedures of direct observation. Upon determining the characteristics of the observable motion, the properties of the second motion may be extrapolated from the established relationship.

Process skills developed by the exercise would appear to be experimenting, controlling variables, and interpreting data.

(Student) The laboratory technique employed in determination of the frequency of a tuning fork is one of using the characteristic of one motion as a basis for determining the properties of a second motion. With the use of the *vibrograph,* tracings of styli attached to a pendulum and a tuning fork are used to determine the ratio of their frequencies. After finding the *frequency* of the pendulum this ratio allows the experimenter to determine the frequency of the tuning fork.

The method of employing tracings for examination of motion is common in science and technology. Instruments such as the seismograph, motor tachometers, and industrial recorders are examples of putting this procedure to work.

PROCEDURE

Over a burning candle (burning camphor in an evaporating dish will also provide smoke to blacken the glass plate) hold a glass plate so as to allow the unburned carbon to coat it. (Note: Holding the plate too close to the flame will cause the plate to crack.) Then place the glass plate on the sliding cradle of the vibrograph, and adjust the tuning fork and pendulum so that the styli are lightly touching the plate. Set the pendulum and tuning fork in motion (some vibrographs are self activating) and quickly draw the glass plate forward under the oscillating styli. The tracing should appear as in Figure 25-1.

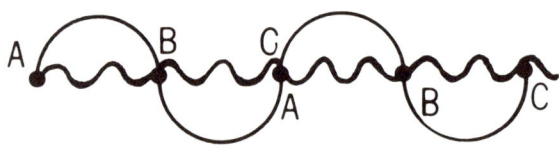

FIGURE 25-1

Show the plate to your instructor before proceeding.
1. Which of the wavelengths was traced by the pendulum? AC (Figure 25-1) is one wavelength and represents one oscillation of the pendulum.
2. How many wavelengths representing oscillations of the tuning fork fall within one wavelength of the pendulum?

(Note to teacher: Some students have difficulty obtaining a trac-

Frequency of a Tuning Fork

ing which shows a full wavelength of the pendulum. It is possible to use plates upon which one-half of a wave is visible to obtain a ratio.)

3. How may the relationship between the frequency of the pendulum and the frequency of the tuning fork be stated?

To determine the frequency of the pendulum start it vibrating and count the number of vibrations that occur in one minute. The pendulum must swing to and fro and return to its original position to constitute one complete vibration.

4. Frequency is generally stated as vibrations per second. Knowing the number of vibrations of the pendulum which occur in one minute how may the frequency in vibrations per second be determined?
5. Can you now determine the frequency of the tuning fork in vibrations per second?
6. In what ways would the speed with which the glass plate is drawn under the styli affect the results of this exercise?
7. How would shortening the length of the pendulum affect the data from the exercise?

DISCUSSION OF OUTCOMES

1. The longer wave was traced by the pendulum.
2. Student answer.
3. Student answer. The relationship may be stated as a numerical ratio.
4. The number of vibrations per second may be found by dividing by 60.
5. The frequency of the tuning fork may be determined by multiplying the frequency of the pendulum by the experimentally determined ratio.
6. The ratio of the frequencies would remain the same. The only effect of speeding up the movement of the glass plate would be to elongate both sets of tracings.
7. A shorter pendulum would vibrate at a faster rate thereby changing the ratio of the frequencies.

EXERCISE 26
DETERMINATION OF THE WAVELENGTH
AND VELOCITY OF SOUND BY RESONANCE

APPARATUS

Resonance Apparatus, or hydrometer jar and resonance tube about 1½" x 18"; tuning fork (256 v.p.s. or higher) for use with the short resonance tube; meter stick; centigrade thermometer.

(Teacher) The concept of resonance is fundamental to the study of sound and has many applications in the fields of electrical circuitry and electronics. In this exercise the student is exposed to physical evidence of this phenomenon. Patterns of destructive and constructive interference become meaningful as the experimenter is exposed to the audible evidence of their existence. Besides experimenting and controlling variables the process skill of making inductive hypotheses is exercised.

(Student) *Resonance* is the reinforcement of sound produced by the uniting of a reflected sound wave with a direct sound wave. This reinforcement occurs only when the direct and reflected waves have the same natural frequency or are multiples of one another. Sound in air is produced by the alternate compression and rarefaction of the air caused by a vibrating body. The compression wave formed at A (see Figure 26-1) will be directed downward by the downward vibrating tuning fork. The wave will strike the surface of the water and rebound to combine with the wave being then formed at the tuning fork. If the length of the closed column (1) is one-fourth the wavelength of the waves formed by the vibrating body, the rebounded wave will combine with the newly produced wave being pushed upward at C in such a way as to reinforce it. Resonance is then said to occur, and the sound produced is audibly louder.

The Wavelength and Velocity of Sound

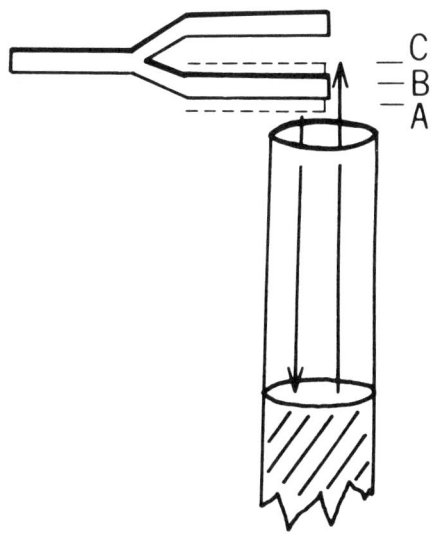

FIGURE 26-1

PROCEDURE

Part A—Finding the Wavelength of Sound

Fill a resonance apparatus or hydrometer jar three-fourths full of water. Hold a vibrating tuning fork over the resonance tube with the fork close to the mouth of the tube. (Caution: Touching the glass cylinder with a vibrating tuning fork may cause damage to the glass.) Increase the length of the air column in the resonance tube until the intensity of the tone increases noticeably. Carefully adjust the length so that the length of the air column that produces resonance may be determined. Measure from the bottom of the tuning fork to the surface of the water.

1. Can you state the wavelength of the sound being generated by the tuning fork? (You may wish to refer to information contained in the introduction to help you solve the problem.)

For increased accuracy in determining the wavelength of a sound, the wave that is reflected from the inside walls of the tube must be considered. It has been found that adding 0.4 of the inside diameter of the tube to the length of the air column accounts for this phenomena. A more accurate estimate of the wavelength of the tone produced by the tuning fork would then be found by: Wavelength = 4 (L + 0.4d) where L is the measured length of the air column and d is the *inside* diameter of the tube.

2. Determine a more accurate estimate of the wavelength of the sound produced by the tuning fork than was given in item 1.

Part B

Examine Figure 26-1 carefully. Does it seem reasonable to hypothesize that resonance will occur with an air column of some other length?

1. Write an hypothesis which represents your best estimate of a second length of the air column which will cause resonance.

Drop the water level in the resonance tube and listen carefully for evidence of resonance. (If short resonance tubes are used, several may be joined together with tape to provide a single long tube.)

2. Did resonance occur? If so, the new found resonating lengths are what multiples of the original length L?
3. Do you accept or reject your hypothesis?

Part C—Determining the Velocity of Sound in Air

Since sound is a wave phenomena, velocity may be determined by multiplying the frequency (f) times the wave length (w). The frequency of the tuning fork used in Part A is inscribed thereon and the wave length has been determined experimentally.

1. Using this data solve for the velocity of sound.

The accepted value of the velocity of sound is 331.5 meters per second plus an additional .6 meter per second for every degree centigrade of air temperature above 0° C.

2. Determine an accepted value for the velocity of sound corrected for temperature.

Vibrations of Strings 125

3. Find the percent of difference between the experimental value of the velocity of sound and the corrected value.

DISCUSSION OF OUTCOMES

Part A

1. The wavelength should approximate 1.3 meters for a 256 v.p.s. tuning fork. The student answer is obtained by multiplying the length of the resonating column by 4.
2. Student answer determined by: wavelength = 4 (L + 0.4d).

Part B

1. Student hypothesis.
2. Resonance should have occurred at L, 3L and 5L.
3. Student answer.

Part C

1. Student answer. (Approximately 330 meters per second.)
2. The accepted value corrected for temperature = 331.5 + .6 (° C) meters per second.
3. Student answer.

$$\% \text{ difference} = \frac{\text{Accepted value} - \text{experimental value} \times 100}{\text{Accepted Value}}$$

EXERCISE 27—VIBRATIONS OF STRINGS

APPARATUS

1. Sonometer with two wires of different diameters
2. Two tuning forks (C, 128 and C, 256 v.p.s.)
3. Weight hanger
4. Weights (A pail of water or sand and a balance may be used)
5. Meter Stick
6. Two Bridges

(Teacher) Examination of the laws of vibrating strings is a common and very appropriate topic for investigation in a unit on sound. The relationships of length, diameter, tension, and density of vibrating strings to the pitch and quality of the resulting sound are basic to an understanding of stringed instruments. In this exercise the student also comes into contact with the concept of destructive interference of sound waves with the formation of "beats" caused by slightly out of phase sound waves. Skills developed by this exercise are experimenting, interpretation of data and understanding numerical relationships.

(Student) A number of musical instruments depend upon vibrating strings to produce tones. The *pitch,* how high or low a sound is, depends upon the frequency of the vibrating string. How the frequency varies as to length, tension, and diameter of the string is the subject of this experiment.

PROCEDURE

Part A—The Effect of Length

Set up the sonometer, as in Figure 27-1, with the bridges about 50 cm. apart and enough weight to cause the string to vibrate at approximately 128 v.p.s. Strike a low C (128 v.p.s.) tuning fork, place the shank on the sounding board of the sonometer and add weights until the plucked string has the same pitch as the tuning fork. As the pitch of the string approaches that of the tuning fork pulsations of sound, or beats, are heard. When the string is carefully tuned by adding weights these beats disappear. After the string is tuned to 128 v.p.s. measure the distance between the tops of bridges A and B.

1. What is the length in cm. of the vibrating string with a frequency of 128 v.p.s.

Now with the same tension on the string, sound the middle C (256 v.p.s.) tuning fork and adjust bridge A until the string vibrates at the same frequency as the tuning fork.

2. What is the length of the vibrating string with a frequency of 256 v.p.s.?

Vibrations of Strings

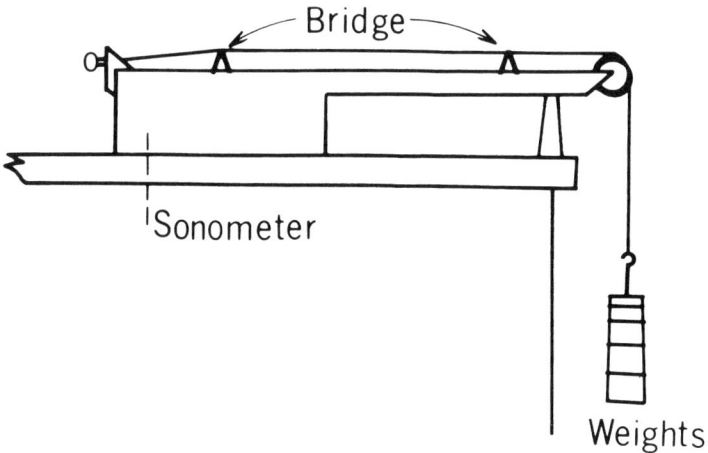

FIGURE 27-1

3. Make a statement as to your conclusion of the relationship of length of a vibrating string to its frequency.
4. If the velocity of a sound emitted by a vibrating string equals its frequency times its wavelength ($V = f\lambda$) how are the frequency and wavelength related for a velocity which is a constant?

Part B—The Effect of Tension
on the Frequency of a Vibrating String

Move bridge A back to where the string is tuned to 128 v.p.s. Record the weight which represents the tension on the string. Now increase the tension by adding weights to the weight hanger until the string is tuned to 256 v.p.s. (Do no put your feet under the weight hanger. A broken string will allow the weights to drop abruptly.) Record the amount of weight (tension) with the string vibrating at 256 v.p.s.

    1. The tension on the string which produced 256 v.p.s. is how many times the tension that produced 128 v.p.s.?
    2. Write a statement that best describes the numerical rela-

tionship which exists between the frequency and tension of vibrating string.

Part C

If a wire with different diameter than that used in parts A and B is available, set it up on the sonometer with the same length and tension as was used to tune the string in Part A to 128 v.p.s. Now pluck the string in the center.

1. How does the pitch differ from 128 v.p.s.?

Using a micrometer measure the diameters of both strings. If the pitch of the new string could be determined accurately by some means the exact numerical relationship between the pitch and diameter of a string might be accurately determined.

2. Can you write a statement that *generally* describes the relationship between the pitch of a string and its diameter?

DISCUSSION OF OUTCOMES

Part A

1. Student answer.
2. Student answer should approximate one-half of item 1.
3. The frequency of a vibrating string is inversely proportional to its length.
4. The frequency and wavelength are inversely related to one another.

Part B

1. Approximately 4.
2. The frequency (pitch) of a vibrating string is directly proportional to the square root of the tension on it.

Part C

1. Student answer.
2. The pitch of a string is inversely related to its diameter.

Intensity of a Light Source

EXERCISE 28—INTENSITY OF A LIGHT SOURCE

APPARATUS
1. One standard lamp (when not available, follow the procedures outlined below)
2. A lamp of unknown candlepower
3. A translucent screen (an oil spot on paper will do)
4. Meter Stick

(Teacher) This exercise serves to illustrate the relationship between the intensity of a light source and its illumination upon surfaces at various distances. A procedure for determining the intensity of a light source is also demonstrated. In this exercise the skills of understanding numerical relationships, understanding space relationships, experimenting, and formulating hypotheses are developed.

(Student) The intensity of a light source is defined as the rate at which it radiates light energy and is usually rated in candlepower. Illumination is a term that defines the amount of light energy that is falling upon a surface—how brightly the surface is lighted.

PROCEDURE

Part A

1. What is your best guess as to how the illumination of a surface is affected by the intensity of the light source which is providing the light?
2. What is your best guess on how distance from the source affects the amount of light energy (flux) that falls upon a surface?

To aid in formulating a more exact hypothesis for item 2, complete the following exercise (see Figure 28-1). Draw a one inch square and locate the center of the square at the intersection of its diagonals. Think of the center of the square as a light source and the surface of the square as being illuminated by the light source. Now extend each ray (0a, 0b, 0c, 0d) so that its length is doubled and connect the end points of the rays to form a second larger square.

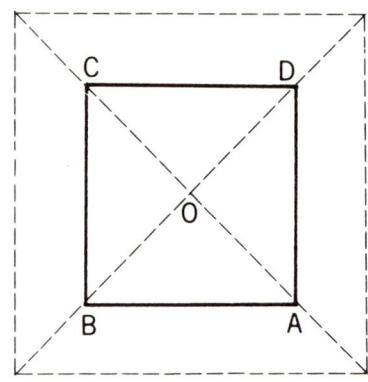

FIGURE 28-1

3. How does the area of the larger square compare to the area of the smaller square?
4. The light flux being radiated from 0 must cover how many times as much area? The illumination of the larger surface will be what fraction of the illumination of the smaller?
5. When the distance from the light source is doubled the illumination on a surface is _____ as great.

In conclusion it may be seen that the illumination on a surface is directly proportional to the intensity of the light source and inversely proportional to the square of the distance from the source. This may be expressed by $E = \dfrac{I}{S^2}$ where E = illumination (in lumens/ft.^{2}), I = intensity (in foot-candles), and S = distance in feet.

Intensity of a Light Source

Part B

To determine the candle power of a lamp of unknown intensity, place the lamp at one end of a meter stick and lamp of known candle power at the other end. If no standard lamp is available, assume a figure using as a guide a 40 watt bulb which is rated at about 35 candle power or a 100 watt bulb at about 135 candle power. Now move a translucent screen between the two light sources until both sides have the same appearance as in Figure 28-2. At this point the illumination of both sides of the screen is equal.

If: E (Standard Bulb) = E (Bulb of Unknown Intensity)

Then: $\dfrac{I}{S^2}$ (Standard Bulb) = $\dfrac{I}{S^2}$ (Bulb of Unknown Intensity)

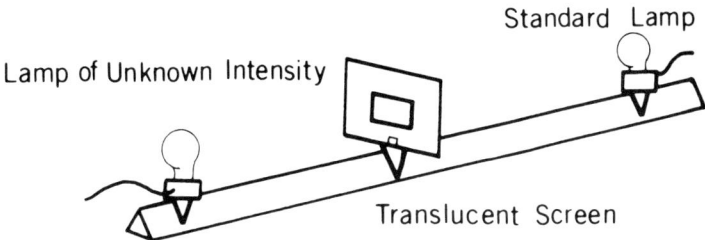

FIGURE 28-2

1. If the distances of the illuminated screen from the two light sources are measured, what is the only unknown factor in the above expression?
2. Determine the intensity of the bulb of unknown candlepower.

Part C

Repeat Part B changing the distance of the light sources from the illuminated screen.

1. How does this result compare with the intensity of the unknown bulb in Part B?

DISCUSSION OF OUTCOMES

Part A

1. Most students should state that the illumination of a surface and the intensity of the light source are directly proportional.
2. Most students will state that the illumination of a surface is inversely related to its distance from the light source.
3. Four times as great.
4. Four times the area. The illumination on the larger square would be one-fourth as great.
5. One-fourth.

Part B

1. The only unknown factor is the intensity of the bulb to be tested.
2. Student answer.

Part C

1. The bulb of unknown intensity should receive about the same foot-candle rating in each case.

EXERCISE 29—REFLECTIONS FROM A PLANE SURFACE

APPARATUS

1. Plane Mirror
2. Ruler
3. Wood Block
4. Pins
5. Protractor

(Teacher) In this exercise the student is acquainted with the laws of reflection. He should come to understand the concept of a virtual image and the orientation and position of that type of reflection. The process dealt with most advantageously in this exercise is that of interpreting data which is obtained through graphic construction.

(Student) Images formed by plane surfaces are the result of light striking a reflecting surface and rebounding to the eye of the observer. These light rays follow some definite patterns of behavior which will be demonstrated by this exercise. Several terms dealing with the phenomenon of light reflection should be understood. The incident ray is the light ray which travels from an object to the reflecting surface. The reflected ray is the light ray which travels from the reflecting surface to the observer's eye. A *normal* is an imaginary line projecting at right angles from the reflecting surface at the point of reflection.

PROCEDURE

Part A

Look at the pictorial illustration, Figure 29-1. The observer at A is observing the image of the object.

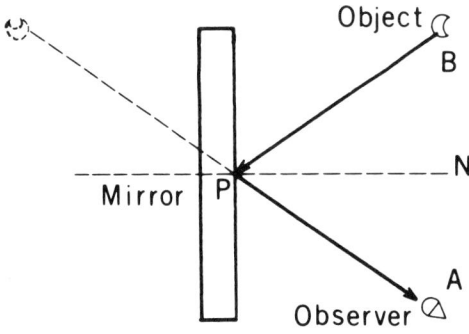

FIGURE 29-1

1. From the definitions provided above can you identify the incident ray? the reflected ray? the normal?

2. According to the illustration where does the image appear to the observer to be located?

Part B

To determine the relationship of the angle of incidence to the angle of reflection, support a plane mirror in the approximate center of a sheet of paper. In front of the mirror draw any point A to be used as an object. Place a straight pin at A. From any position left of A, sight along a ruler edge at the image of A and draw a line AP up to the mirror. The light that formed the image must have traveled from the pin along line AP, and was then reflected at P back to the observer 0. Draw line AP. Scribe a line MN along the front of the mirror, remove it and construct a line P perpendicular to CD as in Figure 29-2. The line is the *normal*.

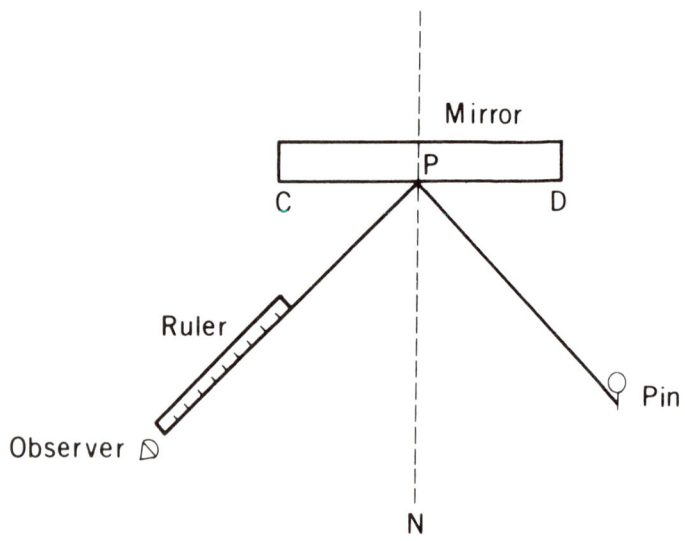

FIGURE 29-2

1. Which angle would be considered the angle of incidence? The angle of reflection?

Reflections from a Plane Surface

2. Measure the angles with a protractor. How do they compare?

Part C

To locate the image of an object in a plane mirror, again support a plane mirror in the approximate center of a sheet of paper (see Figure 29-3). Draw triangle ABC in front of the mirror. Place a pin at point A on the triangle and from some position to the left of point A, sight along a ruler at the image and draw line a_1. (See Figure 29-3.) Now move the ruler to the right of point A and after sighting at the image, draw line a_2. Place pins succeedingly at points B and C and repeat the procedure used in establishing lines a_1 and a_2 to establish b_1 and b_2 and c_1 and c_2. Now remove the mirror and extend lines a_1 and a_2 until they intersect. Do the same to lines b_1 and b_2 and c_1 and c_2. With a straight edge join these three points of intersection. The image of the original triangle is thus established.

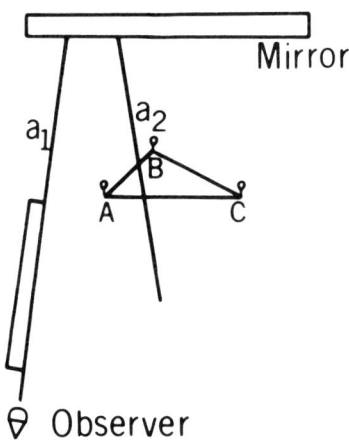

FIGURE 29-3

1. Generally, how does the image triangle compare in size to triangle ABC?

2. How does the distance from the mirror to the object compare?
3. Is the image in the same position or reversed in relation to the triangle ABC?
4. Did the image of triangle ABC actually reside behind the mirror or was it a mental projection?

Images that result from the actual intersection of light rays are called real, while images that are mental projections of observed rays are termed virtual.

5. Into which class do the images formed by a plane mirror fall?

DISCUSSION OF OUTCOMES

Part A

1. Incident Ray—BP, Reflected Ray—AP, Normal—NP
2. The image appears to be located behind the mirror.

Part B

1. Angle of incidence—APN; Angle of reflection—BPN
2. Student answer. The angles should be equal.

Part C

1. The image and object triangles should be approximately the same size.
2. The distance of the image behind the mirror should be equal to the distance of the object in front of the mirror.
3. The image is reversed.
4. The image was a mental projection.
5. The images formed by a plane mirror are virtual.

Images in a Curved Mirror

EXERCISE 30—IMAGES IN A CURVED MIRROR

APPARATUS
1. Convex Mirror
2. Concave Mirror
3. Light Source (or candle)
4. Two Meter Sticks
5. Cardboard Screen
6. Clay or holders for screen and light source

(Teacher) In few exercises in the high school physics laboratory are the textbook descriptions of a phenomena made real more simply than with the exercises involving curved mirrors. The skills of observing and experimenting are the processes which are chiefly utilized in parts A and B. Part C requires the student experimenter to create inductive hypotheses—to generalize on his observations. Some students may require help in achieving understanding of the results of their observations, especially those involving virtual images. It is appropriate to resort to ray diagrams to aid them in this endeavor.

(Student) The curved mirrors which will be used in this experiment may be considered a part of a sphere. If the reflecting surface is the outside of the sphere it is *convex;* if the reflecting surface is the inside of the sphere it is *concave.*

The *principal axis* is a line which passes through the *center of curvature* (C) and the center of the mirror (A). A straight line drawn in any direction from a *center of curvature* will always intercept the mirror's surface at right angles. The principle focus (f) is the point on the principle axis through which light rays approaching the mirror parallel to the principle axis will be reflected (see Figure 30-1). (Some distortion called aberration occurs for large angular sections of spherical mirrors.) The image formed by a concave mirror will vary as the distance of the object changes.

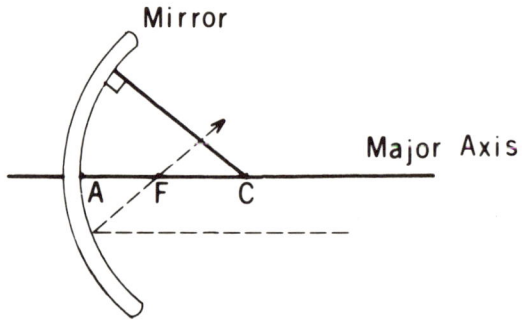

FIGURE 30-1

PROCEDURE

Part A

To determine the focal length of a concave mirror, mount it on a meter stick, hold a light source (a lighted candle will do) close to the mirror and slowly move it back until the image begins to blur (see Figure 30-2). The focal length is the distance between the mirror and the light.
1. What is the focal length?
2. Why is there no image with the light source at f?

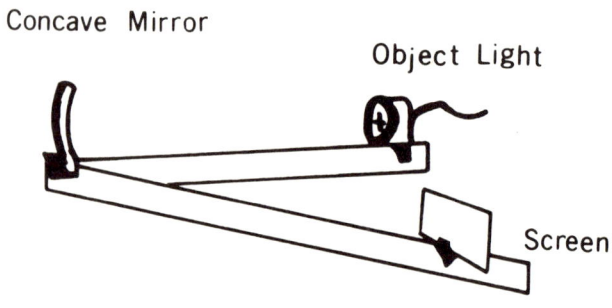

FIGURE 30-2

Images in a Curved Mirror *139*

In a darkened room place the light source some distance beyond two focal lengths (2f) from the mirror with the apparatus as shown in Figure 30-2. Move the screen back and forth until a clear image appears upon it.

3. Note the image's position (cm. from the mirror).
    a. Is it upright or inverted?
    b. Is it enlarged or diminished?
    c. Is it real or virtual?

Move the light until it is two focal lengths from the mirror, and relocate a clear *image* on the screen. Describe the image.

4. Position (cm. from the mirror).
    a. Upright or inverted?
    b. Enlarged or diminished?
    c. Real or virtual?

Move the light source between f and 2f and readjust the screen, until the image is clear and note:

5. Position (cm. from the mirror).
    a. Upright or inverted?
    b. Enlarged or diminished?
    c. Real or virtual?
6. As an object light source moves closer to the mirror its image becomes (*larger, smaller*), and is located (*closer to, farther from*) the mirror.
7. When the light source is placed at f what sort of image is there?

Now place the object between the mirror and f. Describe the image in the mirror.

8. Enlarged or diminished?
9. Upright or inverted?
10. Can you get an image on the screen?
11. The image must be (*real, virtual*).

Part B

To study the image of a convex mirror replace the concave mirror used in Part A with a convex mirror. With the light source at some reasonable distance try to locate an image on the screen.

1. Can you cause an image to form on the screen?

2. The image must be (*real, virtual*).
3. Looking at the image in the mriror it is (*larger, smaller*), (*upright, inverted*), and as the object is moved closer to the mirror the image becomes (*larger, smaller*).

Part C

From your data attempt to make the following generalizations.
1. What kind(s) of mirror(s) under what condition(s) of object distance(s) form virtual images?
2. What kind(s) of mirror(s) under what condition(s) of object distance(s) form real images?

DISCUSSION OF OUTCOMES

Part A

1. Student answer.
2. All light rays emanating from point f strike the spherical mirror at right angles and are reflected from the mirror parallel to one another.
3. Student answer. The image should appear between f and 2f.
    a. Inverted
    b. Diminished
    c. Real
4. The image should appear at 2f.
    a. Inverted
    b. Same Size
    c. Real
5. The image should appear beyond 2f.
    a. Inverted
    b. Enlarged
    c. Real
6. Larger, farther from.
7. No image or extremely blurred.
8. Enlarged.
9. Upright.

Refraction of Light 141

    10. No image will appear on the screen.
    11. Virtual.

Part B

    1. No. No image will form on a screen.
    2. Virtual.
    3. Smaller, Upright, Smaller.

Part C

    1. Plane and convex mirrors under all conditions form virtual images. Concave mirrors form virtual images when the object is placed between the mirror and one f.
    2. Concave mirrors form real images when the object is placed beyond one f.

EXERCISE 31—REFRACTION OF LIGHT

APPARATUS

    1. Thick rectangular glass
    2. Triangular prism
    3. Metric rules
    4. Protractor

(Teacher) Refraction is a phenomenon which is readily observed with simple apparatus in the high school physics laboratory. The effects of refraction are also commonly observable in the students' environment and the principle has many applications to devices involving optics and sound. The process skills required for this exercise appear to be observing and experimenting. In Part D the student is required to apply the understanding he has acquired in the previous portions of the exercise.

(Student) Although light waves travel in straight lines and at constant speeds in any uniform medium, their speed varies in different media. As the light wave approaches the air-glass boundary at A in Figure 31-1, part of the wave front which is composed of light energy enters the glass first and is slowed down while the rest of the wave front continues at its faster rate. This results in a change of direction of the wavefront toward line N which is normal to the boundary. As the wavefront emerges from the glass at B the portion that enters the air first speeds up while the rest of the wavefront continues at the slower rate. The result is a change of direction away from the normal. The *index of refraction* is a constant which describes the amount of bending of light at the boundary of a substance. In this exercise you will determine the index of refraction with several procedures.

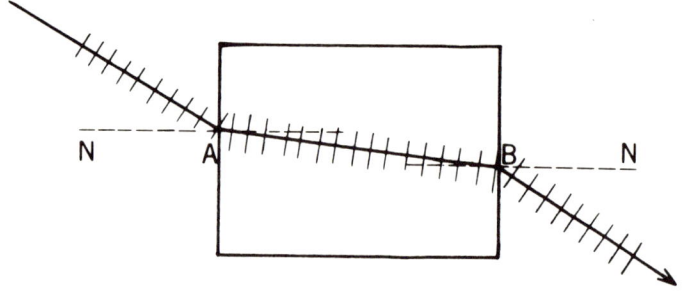

FIGURE 31-1

PROCEDURE

Part A

To determine the path of a light ray through a rectangular glass plate place the plate in the approximate center of a piece of plain paper and trace its outline. Draw line AB so that it strikes the edge of the plate obliquely. (See Figure 31-2.) Place a ruler on the opposite side of the glass plate so that the edge of the ruler appears as a continuation of line AB when sighting along AB through the plate. Draw line CD, remove the plate and connect points C and B with a straight line. Erect at C a line perpendicular to the edge of the glass. This is the normal. Measure the angle of incidence and the angle of refraction with a protractor.

Refraction of Light

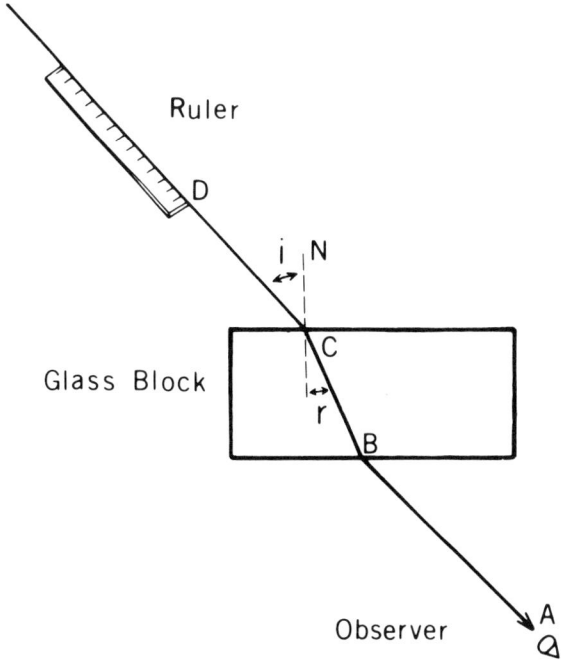

FIGURE 31-2

1. The light ray entering the glass at C was bent (*away from, toward*) the normal.

   The index of refraction may be found by: $\dfrac{\sin i}{\sin r}$

2. What is the index of refraction for this air-glass boundary?

Part B

Repeat Part A using a triangular prism (see Figure 31-3).
1. At point C what is the angle of incidence?
2. The angle of refraction?
3. Compute the index of refraction.

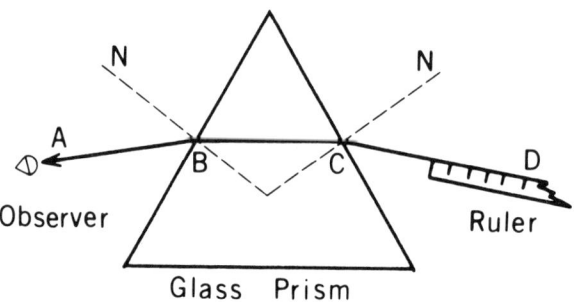

FIGURE 31-3

At B the light ray is passing from a more dense to a less dense medium.

4. The angle of incidence is (*larger, smaller*) than the angle of refraction.
5. The light wave is bent (*towards, away from*) the normal.

Part C

To determine the index of refraction by a second procedure first locate in some reference book the speed of light in a vacuum and the speed of light in flint glass. (The optical density of flint glass is a good approximation of the optical density of most laboratory prisms).

1. Now find the index of refraction by the following relationship:

$$\text{index of refraction} = \frac{\text{speed of light in a vacuum}}{\text{speed of light in a substance}}$$

2. How does this compare with the previously determined values for the index of refraction for glass?

Part D

With what you know about refraction, can you explain the path of the light ray in Figure 31-4?

Refraction of Light

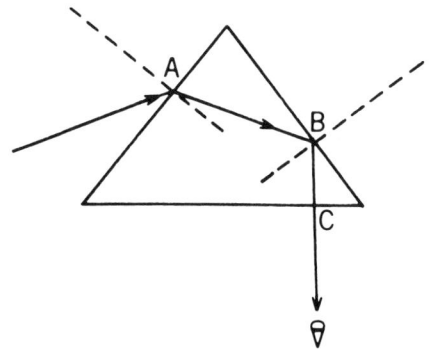

FIGURE 31-4

DISCUSSION OF OUTCOMES

Part A

1. Toward.
2. Student answer.

Part B

1. Student answer (by measurement).
2. Student answer (by measurement).
3. Student answer.
4. Smaller.
5. Away from.

Part C

1. Student answer.
2. The indices of refraction as determined by the two separate procedures should compare favorably.

Part D

At point A the light is bent toward the normal. At point B the light is bent away from the normal to the extent that it does not exit the glass but assumes the path toward C. At C the light is bent away from the normal.

EXERCISE 32—REFRACTION IN A CONVEX LENS

APPARATUS
1. Double convex lens (10 to 20 cm. focal length)
2. Lens holder or clay
3. Meter Stick
4. Screen
5. Light source (or candle)

(Teacher) Images formed by the refraction of light passing through a convex lens have long been a source of intellectual curiosity. The applications of the lens are many and varied. In Exercise 32 the student investigator may observe the effect of lens-object distance upon the projected images. The process skills thought to be developed by the activity are observing and experimenting.

(Student) A lens is convex when it is thicker through the middle than at the edges. The refraction of light through a convex lens is such that it produces a variety of images and is very useful in the manufacture of microscopes, telescopes, cameras, and many other types of optical equipment.

Any light wave passing through the optical center (O) of the lens in Figure 32-1 is not refracted. Any light ray arriving at the lens parallel to the principal axis will be refracted through (f), the principal focus. The focal length of the lens is the distance between O and f. The center of curvature (C) is at the geometric center of the arc forming one side of the lens.

PROCEDURE

Part A

To determine the focal length of a convex lens mount it and a paper screen upon a meter stick. Allow the light from a distant

Refraction in a Convex Lens

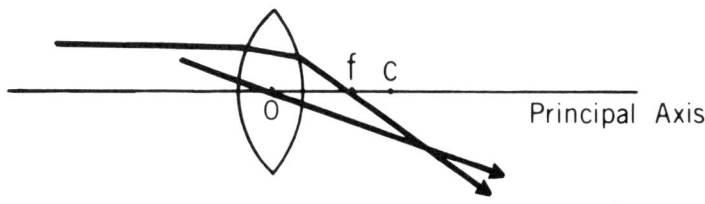

FIGURE 32-1

object outside the window to pass through the lens and form an image upon the screen. Adjust the screen until the image is clear and measure the distance from the lens to the screen. Pick other distant objects and do two more trials.

1. What is the average focal length (f)?

Part B

To study the real images formed by the convex lens mount the lens, a light source, and a screen upon a meter stick as in Figure 32-2. Move the light source to some position beyond two focal lengths from the lens. Move the screen until an image is in sharp focus.

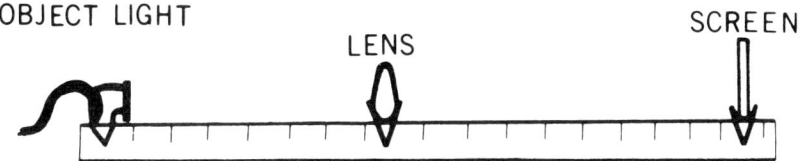

FIGURE 32-2

1. The image for an object beyond 2f is:
    a. real, virtual
    b. enlarged, diminished, the same size
    c. upright, inverted
2. The position of the image is (*less than, more than, the same as*) one focal length from the lens.

Now put the light source at 2f and locate the image.
3. The image for an object at 2f is:
   a. real, virtual
   b. enlarged, diminished, the same size
   c. upright, inverted
4. The position of the image is about (one, two, three) focal lengths from the lens.

Now put the light source between f and 2f and locate the image.
5. The image for an object located between f and 2f is:
   a. real, virtual
   b. enlarged, diminished, the same size
   c. upright, inverted
6. The position of the image is (*less than, the same as, more than*) two focal lengths from the lens.
7. With the light source at f can an image be seen upon the screen?
8. As an object moves closer to a convex lens from a distance its image becomes:
   a. *smaller, larger.*

   With the object at 2f the image is:
   b. (*smaller than, larger than, the same size as*) the object.

   As the object is moved past 2f and closer to f the images become:
   c. *smaller, larger,* until the object is at f and no image may be found on the screen.

Part C

1. With an object less than one f from the lens, may an image be located upon a screen? If there is no real image it is said to be virtual. Draw the lens slowly back from the object until it begins to blur.
2. The image appears to become (*larger, smaller*), as the distance increases. Measure the distance between the object and the lens. This is the focal length of the lens.
3. How does this f compare with f as found in Part A?

The Visible Spectrum 	*149*

DISCUSSION OF OUTCOMES

Part A

1. Student answer.

Part B

1. a. real, b. diminished, c. inverted.
2. more than
3. a. real, b. the same size, c. inverted.
4. two
5. a. real, b. enlarged, c. inverted.
6. more than
7. no image may be seen with the light source at f
8. a. larger, b. the same size as, c. larger.

Part C

1. No image may be located.
2. Larger
3. The focal lengths as determined by the two different procedures should compare favorably.

EXERCISE 33—THE VISIBLE SPECTRUM

APPARATUS

    1. Triangular prism
    2. White, Red, and Blue Reflecting Screens (Colored paper will do)
    3. Red and Blue transparent screens
    4. A light source (sunlight should be used when possible)

(Teacher) This exercise dealing with the visible spectrum is primarily an observational activity. It is designed to aid the student in forming conclusions based upon data which is derived from direct observation. The subject matter is a common natural phenomena, visible light. An objective of the exercise other than developing observational skills is that the student will achieve understanding of the relationship between the wavelength or frequency of light and its color. He should also derive some knowledge of the nature of filtered light and reflected light.

(Student) *Light* is the visible portion of the *electromagnetic spectrum*. It consists of all of the wavelengths between about 4500 and 7000° A. (1 A = 10^{-10} meters). It is possible to separate white light, which consists of all of the wavelengths of the visible spectrum, into separate bands or colors with a triangular glass prism. This separation is possible because each color has a different velocity and is refracted a different amount at the air-glass boundaries. The triangular shape of the prism increases the dispersion of the colors according to the wavelength.

The color of a transparent filter is determined by the wavelengths of light that are transmitted; i.e. a red filter will transmit red light and absorb all others. The color of an opaque substance is determined by the wavelengths that are reflected; i.e. a red object reflects red light and absorbs all others.

PROCEDURE

Part A

To observe the separation of white light into its component colors darken a room and allow a beam of light to pass through a narrow slit cut into a piece of cardboard (if sunlight is available use it) and then through a triangular prism. With the wall of the room or a piece of white paper for a screen observe the spectrum that is formed.

1. Starting at one end of the spectrum name the colors observed.
2. What color is refracted most?
3. How does its wavelength compare with the wavelength of the rest of the colors?
4. How does its frequency compare with the other colors?
5. Can you attempt a generalization concerning the relationship between the wavelength of a color of light and the degree to which it will be refracted by a prism?

The Visible Spectrum 151

Part B

To study the light that passes through a transparent object hold a piece of blue glass or cellophane in the path of the white light that passes through a prism.
1. What colors are now observed?
2. What colors have been eliminated "by the filter"?
3. Do the same with a piece of red glass or cellophane.
4. What colors have been eliminated by the filter?
5. Write a statement which explains the process of filtering light through a transparent substance.

Part C

To study the light reflected from an object let white light pass through a prism and use a piece of blue paper as a screen.
1. What colors are observed as being reflected by the screen?
2. What colors have been eliminated?
3. Do the same using a piece of red paper as a screen. What colors are observed?
4. What colors have been eliminated?
5. Write a statement explaining how the color of an opaque object in white light is determined.

DISCUSSION OF OUTCOMES

Part A

1. The colors of the spectrum are: red, orange, yellow, green, blue, indigo and violet. Usually however, the white light will be resolved into only about four colors.
2. The violet-blue wavelengths are most highly refracted.
3. The violet-blue color bands have the shortest wavelengths of the colors of the visible spectrum.
4. The violet-blue color bands have the greatest frequency of the colors of the visible spectrum.
5. The shorter the wavelength of light, the greater will be the effect of a refracting medium upon it.

Part B

1. Usually only violet and blue.
2. All other colors of the visible spectrum have been eliminated.
3. Usually only red and orange.
4. All other colors of the visible spectrum have been eliminated.
5. The color of an object viewed through a transparent screen is the result of the colors that are permitted to be transmitted by the screen, all others being absorbed by the screen.

Part C

1. Violet-blue.
2. All other colors of the visible spectrum have been absorbed by the screen.
3. Red-orange.
4. All other colors of the visible spectrum have been absorbed by the screen.
5. The color of an opaque object in white light is determined by the various wavelengths of light that are reflected by the object.

EXERCISE 34—WAVES IN A RIPPLE TANK

APPARATUS

Ripple tank and accessories. If no commercial ripple tank is available a very usable unit may be assembled from materials readily found in most homes. A shallow baking dish will serve as a tank. A vibrator may be fashioned from a vibrating motor or a rotating motor with an offset shaft. (Such motors are found in electric shavers, toy automobiles, etc.) Required accessories may be fashioned from large lenses from old optical equipment, microscope slides, and bent sheet metal.

Waves in a Ripple Tank

(Teacher) In this exercise with the ripple tank the student will observe some of the characteristics of wave phenomena. From his observations of wave refraction, diffraction, and interference, he will be asked to infer relationships for the less visible manifestations of the phenomena that occur in nature. It is thought the learning which takes place during this exercise will help the student to better understand wave characteristics in his studies of sound, light, and electronics. The process skills which are strengthened are primarily those of observing and inferring.

(Student) The ripple tank is a device for demonstrating the generation and propogation of transverse waves. The properties of the water waves as visibly demonstrated may also be ascribed to other wave phenomena such as sound and light. What is learned through direct observation about such wave characteristics as interference, reflection, diffraction, and refraction will help you to better understand the behavior of wave phenomena that occur in less visible forms.

PROCEDURE

Part A

Set the ripple tank on a level surface. Place a single bent wire generator in the agitator arm, and after filling the tank with water to within one inch of the edge, adjust the bent wire generator until it just touches the surface of the water. Connect an appropriate source of current to the input binding posts and adjust the vibrator and strobe (when available) control knobs to the slow positions and darken the room. On units which contain a strobe adjust the speed of the strobe until the waves in the tank appear to be moving slowly or stop. This image represents wave propogation emanating from a point source.
1. Draw a sketch that shows this.
2. Can you explain how the strobe effect of a rotating slotted wheel serves to "stop" the motion of the waves?

Part B

To study the interference patterns of waves emanating from

two point sources, place two single wire point sources in the agitator at the same time and regulate the wave and strobe control knobs until a stationary wave interference pattern is observed.
1. Draw a sketch that shows this pattern.
2. Describe any difference in amplitude between the waves that result from the interference and the normal waves.

Part C

To illustrate wave refraction, install the straight wave generator and place a thick lens or microscope slides in the water to reduce the depth. The bending of a wave front due to change in velocity is analogous to refraction.
1. Draw a sketch to show the refracted wave pattern.
2. Describe a situation where light waves produce this same condition.

Part D

The phenomenon of diffraction may be explained by Huygens' principle: Every point on a wave front acts as though it were a center of disturbance sending out disturbances of its own to form a new wave front. With the straight wave generator installed, place two reflectors close together so that waves expand out in a circular fashion from the space between them. Observe the wave front which emanates from the narrow slit.
1. Draw a sketch showing the diffraction of waves.
2. Can you describe a situation wherein sound waves appear to be diffracted?

Part E

Reflected waves form interference patterns with oncoming waves when they rebound from a surface. Place one or two reflectors in the tank and generate first straight, then point source waves. Adjust the angle of incidence and the distance to the agitator for maximum effect. Draw a sketch for both the:
1. straight and
2. point source waves.
3. Look at the sketch of interference patterns of Part E.

Waves in a Ripple Tank

Describe a situation concerning sound waves where this condition is observed.
4. For the water waves emanating from a point source how does the energy with which the wave strikes a unit surface vary with the distance from the source?

DISCUSSION OF OUTCOMES

Part A

1. Student sketch.
2. The motion of the slotted wheel is such that a slot rotates into position for observation each time a new wave passes, giving the appearance of a stationary wave.

Part B

1. Student sketch.

Part C

1. Student sketch.
2. Refraction occurs where light passes at an oblique angle from one transmitting medium to a second transmitting medium such as from air to glass.

Part D

1. Student sketch.
2. Sound may be heard around corners of obstructions.

Part E

1. Student sketch.
2. Student sketch.
3. "Dead spots" in auditoriums result from the destructive interference which occurs when reflected waves intermingle with source waves.
4. Since water waves are broadcast in circular patterns from a point source increasing the distance of a reflecting sur-

face from the source will cause a proportional decrease in the energy of an equal section of a wave front.

EXERCISE 35
OPTIONAL PROBLEMS TO TEST LABORATORY SKILLS

(Teacher) In Exercise 35 students are again asked to meet the challenge of solving a problem independently. It is important that the teacher reestablish the proper atmosphere and shape student attitudes to derive maximum benefit from the learning situations. The students should be challenged to show their capabilities in working as independent investigators. The teacher will again function as an advisor and consultant, providing direction for the students' efforts, but not structuring their efforts for them.

Problem A is designed to be the simplest of the suggested problems and it is suggested the less able students be influenced to choose it for their optional exercise. All that is required is for the students to arbitrarily choose some lengths for baselines and complete measurements in order to check their results for the employment of the parallax method of estimating distances.

Problem B offers the greatest challenge in terms of theory and experimental design. It is suggested that the more able students be influenced to choose it for their optional problem. Some students may wish to estimate the wavelength of light by diffraction procedures described in most college physics texts. Those who choose the procedure suggested here will need to count the interference bands that occur in a unit of length on the glass plate. Multiplying this number times the length of the glass plate yields the number that would occur over the entire length of the block. After establishing the amount of separation of the glass plates by measuring the fine wire separator the student may divide the magnitude of the separation by the number of striations in the length of the glass plate to find the wavelength.

To deal with Problem C the investigator must devise a stroboscope by cutting slits in a cardboard or sheet metal wheel and providing an axis at the center upon which it may rotate. Then they must synchronize the rotation of the strobe with the motion

Optional Problems

of the moving object so as to produce an apparent stationary position. Next they must determine the rate at which the strobe is rotating and multiply this times the number of slits in the strobe.

Four separate problems and suggestions for their investigation are listed below. It is up to the student to devise and carry out plans for experimentation and to draw reasonable conclusions from the data. If necessary refer to Unit I, Experiment 9, for instructions on writing a laboratory report.

PROCEDURE

Problem A—Measurement of Distance by Parallax

Astronomers use the parallax method to determine the distance to the closer stars. Can you adapt this procedure to distance measurement of more ordinary magnitude?

With the observer in position A, star Y appears to be aligned with Z (Figure 35-1). With the observer at position B and sighting towards Y triangle AYC is formed which has a known side AC. Triangle BDC may be constructed so $\overline{BD}$ and $\overline{DC}$ are known. The similarity of Triangles AYC and DBC and the proportion $\dfrac{AC}{DB} : : \dfrac{AY}{DC}$ may be established. The student should formulate mathematical proofs and conduct experiments to show that this procedure may be used to estimate distances in the magnitude of common surveying measurements. (A few feet to hundreds of feet).

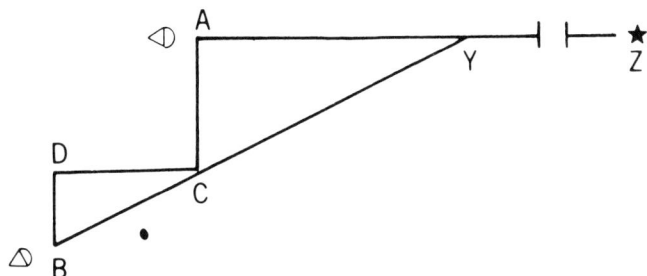

FIGURE 35-1

Problem B—The Estimation of the Wavelength of Light

When a very thin object (a fine wire will do) is used to separate one end of two glass plates, which are bound tightly together by rubber bands, and a source of monochromatic light, such as a mercury vapor lamp, is held above the surface, dark bands appear in the reflected image of the light source. (See Figure 35-2.) These striations result when the light waves reflecting from the surface of the lower plate interfere with those reflected from the upper plate. How much out of phase must the reflected light waves be to produce the dark bands? How is the thickness of the material which is creating the air space related to the number of black bands on the length of the glass plate?

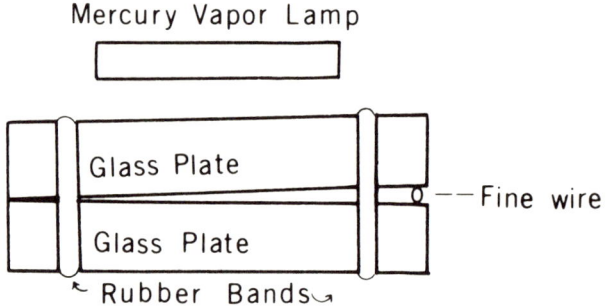

FIGURE 35-2

Can you approximate the wavelength of the light using the procedure suggested above or any other procedure of your own choosing?

Problem C—Determining the Rate of Motion of a Rotating or Vibrating Body

(Student) Why do the wheels of a moving stagecoach appear to be stationary or slowly turning in comparison to the vehicles rate of motion in moving pictures? The answer lies in the strobe effect. When rotating or vibrating objects are viewed at timed inter-

Optional Problems

vals so that the body has returned to the same position on each viewing, the motion appears to be "stopped" to the viewer. Can you determine the rate of motion of several rotating or vibrating objects (tuning forks, color wheel, etc.) and describe your procedures in a report?

Unit V

Static Electricity and Direct Current

The Use of Meters
Static Electricity
Ohm's Law and Resistance
Cells in Series and Parallel
The Electrolytic Cell and the Activity Series
The Lead Storage Cell
Electroplating
Optional Problems to Test Laboratory Skills

Unit V

THE USE OF METERS

In the next weeks you will be using electric meters in laboratory. Since these meters are expensive and easily damaged, it is wise to know something of their construction and the proper methods of use before employing them in experimental work.

Most common meters are of the galvanometer type. In this type meter a small current is allowed to flow through an armature composed of fine wire wrapped around an iron core. With the current thus flowing, the armature becomes an electromagnet which orients itself in an external magnetic field. This orientation results in the movement of the indicator. *The armature of such instruments are wound with very fine wire and the heating effect of any excessive current will damage them.*

Use of the Galvanometer

A two-post galvanometer is a very sensitive instrument and should be used only to detect very small induction voltages such as that resulting from a bar magnet passing through a coil of wire. If a galvanometer is called for and a multipost galvanometer is available, first connect the instrument using the "high range" posts and successively introduce the most sensitive ranges until the one of optimum value is found. *The galvanometer is to be connected in series* in a circuit.

Use of the Ammeter

An ammeter is always connected in series in a circuit. All of the current of the circuit must be carried by the instrument. Most of the current is shunted around the armature, but if the ammeter is overloaded, the armature will carry more than it is designed to carry resulting in damage to the instrument. Before a two-post meter is connected in a current bearing circuit, be sure the magnitude of the current is not beyond the range of the instrument. *Check with your instructor if you have doubts.* If a multipost meter is to be used, start with the most gross range and work successively towards the most sensitive range until the instrument may be read accurately.

Use of the Voltmeter

A voltmeter is always connected in parallel. Resistors in the instrument prevent the armature from being overloaded when used properly. However, damage to the instrument may result if used where it is forced to carry too much current. If a single-range meter is used, be sure the voltage to be measured is not beyond the range of the instrument. *Check with your instructor if you have doubts.* In using a multirange voltmeter, start with the highest range and successively introduce each more-sensitive range until the desired results are obtained.

A common notion of beginning students—that they may limit the amount of current or the heating effect in a circuit by quickly closing and opening a contact key or switch—is false. *Do not use too sensitive a meter and think to prevent damage to it by exposing it for only a short time to the current.* If the instrument is badly overmatched, it will be damaged instantaneously by the too-high current.

All D.C. instruments must be connected with the proper polarity. The negative markings of the source are connected with the negative marking of the meter and positive source to the positive terminal of the meter. Conventionally sources of D.C. current use the color red to designate the positive terminal and a black to designate the negative terminal.

If the precautions contained in the preceding paragraphs are observed, the laboratory equipment will provide you with the accurate data and no damage will result to the instruments. *Study this information carefully.*

Static Electricity

EXERCISE 36—STATIC ELECTRICITY

APPARATUS
1. Glass Rod
2. Bakelite Rod
3. Silk Cloth
4. Wool cloth or cat's fur
5. Pith Ball
6. Silk Thread
7. Stand
8. Electroscope

(Teacher) This exercise employs the use of electroscopes to examine some properties of static electricity. The student is required to observe relationships of charged objects and to draw generalizations from his observations. Hypothesis-making is a skill required for Part C of the exercise. As a result of his work with static electricity the student should come to understand the role of the electron as a transportable charged particle. This understanding is required for subsequent study of current electricity.

(Student) Atoms are composed of many charged particles. Negatively charged electrons are found in the outer regions of the atom and positively charged protons are found in the nucleus. Objects may become negatively charged by gaining electrons and positively charged by losing electrons. Those charges that reside on the surface of objects are known as *static electricity*.

PROCEDURE

Part A

To obtain a *negative charge,* rub a bakelite or hard rubber rod with a piece of wool or fur. Touch the rod to a pith ball which has been suspended from a stand by means of a silk thread. Recharge the rubber rod by again rubbing it with a piece of wool. Bring the rod close to the charged pith ball.

1. What happens?
2. What might be concluded about two negatively charged bodies? Discharge the pith ball by touching it. Rub a glass rod vigorously with a silk cloth to obtain a *positive* charge and touch it to the pith ball.
3. What happens?

(Note: Positive charges are generally weaker and more difficult to sustain than are negative charges.)

4. What might be concluded about two positively charged bodies?

Now again charge the pith ball with a negative charge (bakelite rod and fur) and bring a positive charge (glass rod and silk cloth) close to it.

5. What happens?
6. What might be concluded about dissimilarly charged bodies.
7. If only electrons are transported, how may a positive charge be obtained as on the glass rod?

Part B

To charge a leaf electroscope by contact, rub a rubber rod with a piece of wool and touch the rod to the ball (center rod) of the electroscope.

1. Result?
2. What happens to the leaves of the electroscope? When the rod is removed, touch the knob of the electroscope with your hand.
3. Result?
4. Why does this happen?

Part C

To observe the induction method of charging an electroscope, bring a charged rubber rod close to the knob of the electroscope without touching it.

1. What happens to the leaves of the electroscope? Remove the rod.
2. What happens to the leaves?

Static Electricity

3. Can you offer an hypothesis to explain this observation? Again bring a charged rubber rod close to the knob of the electroscope.
4. The leaves _____ as before, but when a finger touches the knob the leaves (5) _____. Now remove both the rod and the finger.
6. Result?
7. Can you offer an hypothesis to explain this observation?

DISCUSSION OF OUTCOMES

Part A

1. The pith ball is repelled.
2. Negatively charged bodies repel each other.
3. The pith ball is repelled.
4. Positively charged bodies repel each other.
5. The pith ball is attracted to the rod.
6. Bodies with unlike charges attract each other.
7. Positive charges are obtained by the removal of negative charges from atoms which leaves a residual positive charge.

Part B

1. The leaves separate.
2. The leaves remain separated.
3. The leaves collapse to their original positions.
4. The charge on the leaves is drained off through the touching hand.

Part C

1. The leaves separate.
2. The leaves collapse to their original position.
3. The negative charge on the rod repels negative charges of the metal atoms of the electroscope to the leaves causing them to repel each other. Removal of the rod allows the electrons to redistribute themselves uniformly again.

4. Separate.
5. Collapse.
6. The leaves separate.
7. The finger touching the knob of the electroscope provides a path for electrons when a negatively charged rod is held near. The loss of the electrons leaves a residual positive charge on the electroscope and the leaves separate as a result.

EXERCISE 37—OHM'S LAW AND RESISTANCE

APPARATUS

1. D.C. source of current (dry cells, storage battery, generator)
2. Multirange ammeter
3. Voltmeter
4. Insulated connecting wires
5. A resistance wire (A)
6. A resistance wire (B), twice the length and equal in dia. to (A)
7. A resistance wire (C), the same length and twice the dia. of resistor (A)

(Teacher) This is the students' first experience with current electricity. It may be desirable to demonstrate for them the proper use of laboratory meters and remind them to read the section of this unit titled "The Use of Meters." The exercise is designed to help the students establish the interrelationships of current, resistance, and voltage in a D.C. circuit. Parts B and C are devoted to helping the students to gain insight into the conducting properties of copper wire. The process skills required in the exercise would appear to be experimenting and understanding numerical relations.

(Student) The resistance to the flow of electric current through a circuit is measured in *ohms*. The magnitude of the resistance may be computed using the form of Ohm's Law: $R = \dfrac{I}{R}$, where R = the resistance in ohms, E = the electromotive forces

Ohm's Law and Resistance

in volts and I = the current in amperes. The resistance of the resistor shown in Figure 37-1 may be approximated by the voltmeter-ammeter method. That is: $R = \dfrac{\text{reading of the voltmeter}}{\text{reading of the ammeter}}$.

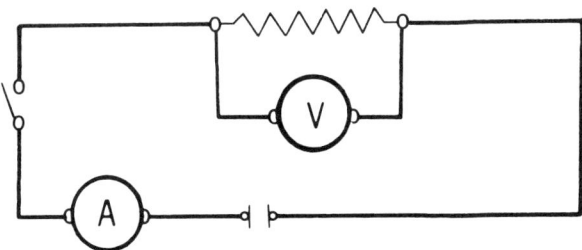

FIGURE 37-1

To determine the effects of length and cross-sectional area upon resistance it is necessary after finding the resistance of resistor A to have available a resistor B of the same diameter, but twice the length, and resistor C of the same length as resistor A, but twice the diameter. (Note: #22 gauge wire has twice the diameter of #28 gauge wire and #24 gauge has approximately twice the diameter of #30 gauge wire. Each decrease of six gauge numbers approximately doubles the diameter of the wire.)

PROCEDURE

Part A

After carefully reading the page entitled "The Use of Electric Meters," proceed to construct the circuit diagramed in Figure 37-1, using the resistor designated as A.

Part B

To find the effect of length of a resistor upon resistance, substitute for resistance A a wire of the same diameter, but twice the length.

1. What is the resistance?
2. How does this resistance compare to resistance A?

Part C

To find the effect of the diameter of a resistor upon resistance, insert a resistor of the same length as resistor A, but of twice the diameter into the circuit of Figure 37-1.
1. What is the resistance?
2. This resistance is what fraction of resistance A?

Complete the following reasoning process to better understand the relationship of the diameter of a wire and its properties as a resistor: The cross-sectional area of a wire is circular in shape.

3. When the diameter of a circle is doubled, what is the effect upon the area of the cross-section?
4. What effect should doubling the diameter of a wire have upon its ability to conduct an electric current?
5. Since $R = \dfrac{V}{I}$ how would the resistance (R) be affected when the current (I) is changed by the factor decided upon in item 4?

DISCUSSION OF OUTCOMES

Part A

1. Student answer.

Part B

1. Student answer.
2. The resistance should be twice as great.

Part C

1. Student answer.
2. Approximately ¼.
3. When the diameter of a circle is doubled, the area becomes 4 times as large.

Cells in Series and Parallel

4. Doubling the diameter of a wire should increase its ability to conduct a current by a factor of 4.
5. Increasing the current by a factor of 4 would mean a reduction in the resistance by the same factor.

EXERCISE 38—CELLS IN SERIES AND PARALLEL

APPARATUS
1. 1-2 ohm resistor or resistance box
2. Three dry cells
3. Voltmeter (0-6 volts)
4. Ammeter (0-30 amps)
5. Ammeter (0-3 amps)

(Teacher) In this exercise the student is exposed to experiences which should cause him to be aware of the differing characteristics of batteries of cells which are connected in series as opposed to those connected in parallel. As he works with Ohm's Law the relationship of current, voltage and resistance should become more clearly fixed in his mind. Although technological considerations are prominent in the design of this activity the nature of the questions causes the student to make inferences, interpret data, make an hypothesis, and to make a test of Ohm's Law.

(Student) A *battery* consists of two or more cells connected together. The cells are joined in *series* (positive to negative) to form a single conducting path for the current through the cells in order to gain maximum voltage, and in parallel (positive to positive and negative to negative) to gain maximum life from the battery.

PROCEDURE

Part A

Connect a voltmeter (0-3 or 0-6 voltrange) across the terminals of a single dry cell.

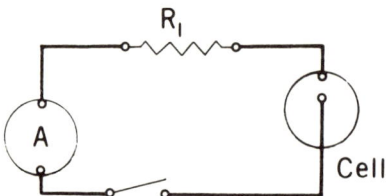

FIGURE 38-1

1. What is the voltage reading?
   Now connect an ammeter (0-30 amps) across the terminals of a single dry cell. (Note: Be sure the ammeter is designed to carry up to 30 amps of current. *Close the circuit for only an instant. Don't drain the dry cell.*)
2. What is the amperage output of the cell?
3. What must be the total resistance of the circuit? (Remember Ohm's Law). Connect a resistance of known value, (1-2 ohms) R_1, with an ammeter (0-3 amps) as in Figure 38-1.
4. What is the ammeter reading?
5. Does your data support Ohm's Law?
6. What are some factors (variables) which you could not control in the exercise?

Part B

Now connect three cells in series as shown in Figure 38-2. Connect a voltmeter (0-6 volts) to the positive terminal of the first cell and the negative terminal of the third. With the resistor R_1, the ammeter (0-3 amps), and the switch all in series close the switch long enough to obtain the voltmeter and ammeter readings.

1. How does the voltage compare with the voltage of a single cell?
2. What is the ammeter reading?
3. The current with three cells in series is approximately _____ times the current of part A because _____.

Cells in Series and Parallel

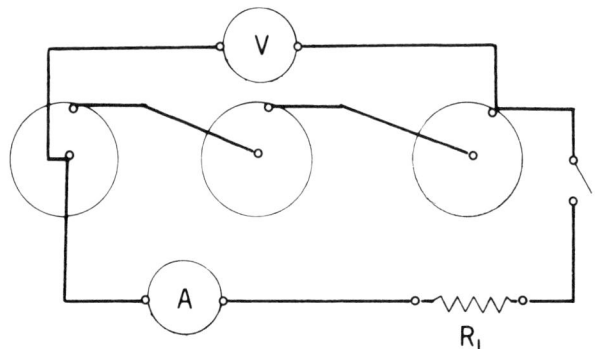

FIGURE 38-2

Part C

Connect three cells in parallel as shown in Figure 38-3. Place a voltmeter across the terminals of the third cell.

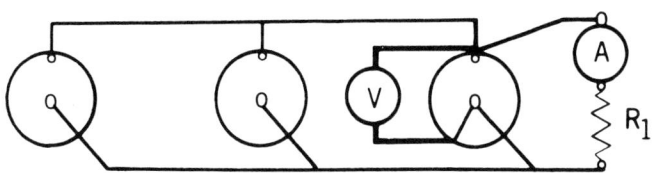

FIGURE 38-3

1. What does it read?
2. How does the voltage with three cells in parallel compare with the voltage of a single cell?

Now connect a switch, resistor R and an ammeter in series to the terminals of the third cell. Close the switch only long enough to obtain an ammeter reading.

3. How does the amperage with the three cells in parallel compare with the amperage in Part A?
4. If the amperage of the circuit in Part C is slightly greater, explain.

5. If connecting cells in parallel only slightly increases the amperage output, can you guess another characteristic of the system which might be considered an advantage?

DISCUSSION OF OUTCOMES

Part A

1. The voltmeter should read approximately 1.5 volts.
2. A fresh dry-cell will yield approximately 25 amps of current.
3. Student answer derived by dividing answer 1 by answer 2.

$$R = \frac{V}{I}$$

4. Student answer.
5. Student data should support Ohn's Law.
6. The student could not control the internal resistance of the cell, the resistance of the connecting wires, and the inaccuracies of the meters.

Part B

1. The voltage of three cells in series is approximately three times as great as a single cell.
2. Student answer.
3. Three times. According to Ohm's Law tripling the voltage should triple the current in the same circuit.

Part C

1. The voltmeter should read approximately 1.5 volts.
2. The voltage is approximately the same.
3. The amperage may be slightly greater.
4. The amperage is slightly greater due to a drop in the internal resistance of the cells connected in parallel.

$$\frac{1}{R_T} = \frac{1}{R_1} + \frac{1}{R_2} + \frac{1}{R_3}$$

The Electrolytic Cell and Activity Series 175

5. Long life. Three cells connected in parallel will supply three times the electrical energy as will a single cell.

EXERCISE 39
THE ELECTROLYTIC CELL AND THE ACTIVITY SERIES

APPARATUS
1. Strips of copper, zinc, aluminum, iron and lead
2. Dilute H_2SO_4 (1:20)
3. Milliammeter
4. Voltmeter (1-3v)
5. Multirange Galvanometer
6. Battery Jar or Beaker
7. Battery Stand
8. Sandpaper or Steel Wool

(Teacher) This exercise dealing with electrolytic activity is often regarded by the student as one of the most rewarding exercises they experience in the study of current electricity. Understanding the electrical properties of materials aids the students in grasping the fundamentals of current flow and potential difference. It also enhances appreciation of the technological applications of the demonstrated principle of the activity series. The process skills involved in the development of the exercise would appear to be observing, inferring, and interpretation of data in Part A, and experimenting, controlling variables, and interpreting data in Part B.

(Student) Any two dissimilar metals immersed in an electrolyte will generate an electric current across a conducting path. The metal atoms with the greatest tendency to yield electrons will comprise the cathode and the metal atoms with the lesser tendency to yield electrons the anode. The potential difference between two metals when immersed in an electrolyte depends upon their position on an activity series that starts with the metal with the greatest electronegativity (electron yielding tendency), and is arranged in descending order of this property. The voltaic cell is the classic example of the electrolytic cell and a knowledge of its properties will lend understanding to the whole concept of electrolytic action.

PROCEDURE

Part A

To construct a voltaic cell, polish copper and zinc strips with steel wool or sandpaper and fix them in a battery stand. Immerse the metal strips in a dilute solution of sulfuric acid, as shown in Figure 39-1. Connect a galvanometer between the electrodes and note the number of galvanometer units that are registered. (Note: Do not use a sensitive galvanometer without placing a resistor in series with the meter.)

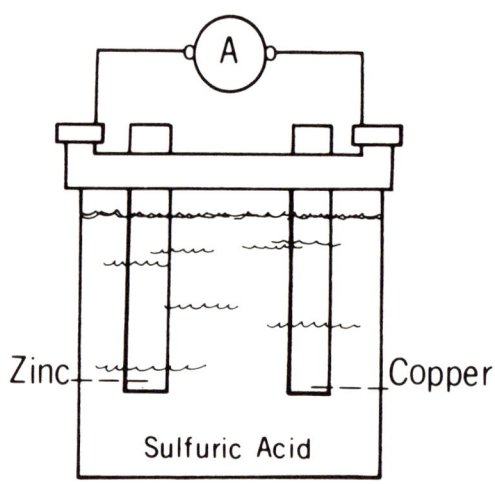

FIGURE 39-1

1. Note the number of galvanometer units that are registered. The zinc plate is the cathode and the copper plate is the anode.
2. Did the indicator deflect toward the cathode or anode post? This is a method of determining the direction of the

The Electrolytic Cell and Activity Series

current which always emerges from the cathode and flows toward the anode. Now attach a milliammeter to the voltaic cell.

3. At the instant the current flows the ammeter reading is _____ ma.

Note the appearance of the electrodes at this instant to be used for a later comparison. Allow the current to flow for about 3 minutes and again read the milliammeter.

4. How does this reading compare with the initial reading? Look at the electrodes again.
5. What could account for the change in the current output of the cell? Now wipe off the electrodes and again immerse them in the acid bath. Attach an ammeter.
6. What happens to the current output as the electrodes are brought closer together?
7. Moved farther apart?
8. As they are slowly raised from the electrolyte? By connecting a low range voltmeter (0-3 volts) to the cell determine the Emf of the cell.
9. Emf = _____ volts.

Part B

Can you determine the proper order of placement on the activity series for Lead, Copper, Zinc, Iron and Aluminum? (Any other metals may be substituted for any of those listed.) Hint: Always polish the metal strips.

1. Why?
2. Make an attempt each time to immerse the same surface area of each metal. Why?
3. Can you name any other variables that should be accounted for in doing the experimentation?

Remember: A galvanometer may be used to indicate the direction of current flow. The current will flow from the more active to the less active metal.

4. List the metals in order of their electronegativity beginning with the most active.

DISCUSSION OF OUTCOMES

Part A

1. Student answer.
2. Student answer.
3. Student answer. The electrodes appear to be clean and free from accumulated bubbles.
4. The milliammeter reading dropped perceptibly.
5. Bubbles of gas appear to have accumulated about the cathode.
6. The current output increases.
7. The current output decreases.
8. The current output decreases.
9. The emf of the cell is approximately 1.1 volts.

Part B

1. Oxide coating and corrosion on the metal strips will shield the metal from the electrolytic solution.
2. The amount of exposed surface area is a variable that affects the amount of current flow.
3. The student might name the distance between electrodes, strength of the electrolyte, and cleanliness of the connectors and contacts.
4. For the metals suggested: Aluminum, Zinc, Iron, Lead, Copper.

EXERCISE 40—THE LEAD STORAGE CELL

APPARATUS

1. Four lead strips
2. D.C. Source
3. Connecting Wires
4. Ammeter (1-20 amps)
5. Battery Jar or Large Beaker
6. Battery Stand
7. Dilute Sulfuric Acid (1:10)
8. Sandpaper or steel wool

The Lead Storage Cell

(Teacher) The lead storage cell differs from the voltaic cell in its ability to be regenerated. For this reason it has been widely adopted for commercial use. This exercise should aid the student in achieving some understanding of the lead storage cell. In Part C the student is required to write and test an hypothesis dealing with the effects of varying the charging time of a cell. Process skills which appear to be required in the exercise are hypothesizing, experimenting, observing and inferring.

(Student) the *lead storage cell* is an electrolytic cell in which lead peroxide (PbO_2) is utilized as the anode and lead (Pb) as the cathode. The electrolyte is dilute sulfuric acid (H_2SO_4) with a specific gravity of about 1.29. The lead cell is a very popular means of delivering D.C. current because its electrodes may be regenerated by *charging*. Charging is accomplished by sending a current through the cell in the direction opposite to the discharging current. The chemical equations that represent reactions during discharge of the cell are as follows:

Cathode: $Pb^0 + SO_4 = PbSO_4 + 2e^=$
Anode: $2H^+ + 2e^- + PbO_2 + H_2SO_4 = PbSO_4 + 2H_2O$

PROCEDURE

Part A

After cleaning two lead strips with steel wool, immerse them in a dilute sulfuric acid bath (1:10) and connect an ammeter between them (see Figure 40-1).
1. Is there a current?
2. Why or why not?

Now remove the ammeter and send a direct current through the cell for about five minutes. Reconnect the ammeter (1 to 20 amps), and find:
3. the maximum current output and
4. The time it takes the cell to discharge completely.

Part B

Attach two clean lead plates at each electrode position so that maximum surface area is exposed to the electrolyte. Again

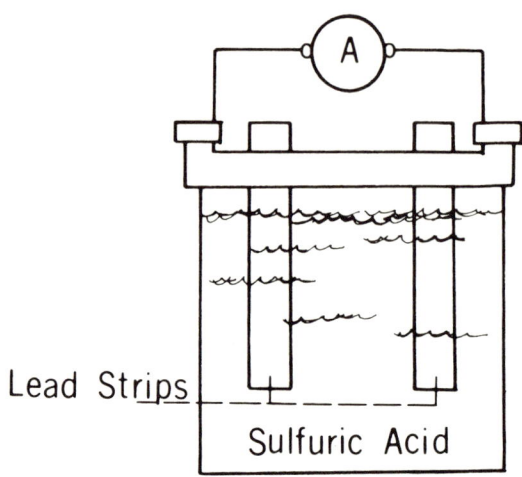

FIGURE 40-1

charge for five minutes with the same D.C. source as in Part A. Find:
1. the maximum current output and
2. the time for complete discharge of the cell.
3. Is the current output higher when the area of the electrodes is greater?
4. Is the discharge time increased with an increase in electrode area?
5. When buying a commercial storage battery, what is one important factor to consider in light of the just completed exercise?
6. Why is the specific gravity of the electrolyte a good standard for determining the state of charge of a lead storage cell?

Part C

How does the time which a charge is applied to a storage cell affect the life of a cell? The data for the time and rate of

The Lead Storage Cell

discharge of Part B is known. Write an hypothesis which predicts the effect upon time and rate of discharge when the same cell is charged for 10 minutes. Test your hypothesis and justify the result.

DISCUSSION OF OUTCOMES

Part A

1. There should be no current.
2. The electrodes are of the same metal; there no potential difference exists between them.
3. Student answer.
4. Student answer.

Part B

1. Student answer.
2. Student answer.
3. The current output is only slightly enhanced by the increase in electrode area.
4. The time of discharge is appreciably affected by an increase in electrode area.
5. The number of plates which is equivalent to a statement of electrode area.
6. As a lead storage cell discharges the sulfuric acid is used up on the chemical reaction, thus reducing the specific gravity of the electrolyte.

Part C

Charging the laboratory cell for ten minutes will not appreciably increase its current output or time of discharge over the five minute charging time. The buildup of PbO_2 at the cathode during charging will at some time reach a point where the efficiency of the chemical reaction is greatly reduced.

EXERCISE 41—ELECTROPLATING

APPARATUS
1. D.C. Source
2. Carbon Rod
3. Copper Strip
4. Steel Wool or Sandpaper
5. Ammeter (0-3 amps)
6. Copper sulfate solution
7. Beaker
8. Battery Stand
9. Connecting Wires
10. Timer

(Teacher) The process of electroplating involves some very basic concepts of atomic structure and current electricity. This exercise should help the student to bridge the abstractions of atomic theory and confirm his notions of the electrical nature of matter. When dealing with electrolytic and electromotive cells the teacher may stress the oneness of science, in this instance the merging of chemistry and physics. The process skills most utilized in this exercise are observing, inferring, hypothesizing, and model bulding.

(Student) *A metal ion* is a *metal atom* which has lost one or more electrons. When the lost electrons are replaced the ion is again converted to an atom. The basic principle of electroplating is to supply electrons to metal ions in solution causing them to be converted to atoms which collect upon the cathode of the cell. At the anode metal atoms give up electrons and become ions in solution, one such atom ionizing for each ion plated at the cathode.

For the cell illustrated in Figure 41-1 the electrode reactions are:

$$\text{Anode:} \quad Cu^0 - 2e^- \rightarrow Cu^{2+}$$
$$\text{Cathode:} \quad Cu^{2+} + 2e^- \rightarrow Cu^0$$

Electroplating

PROCEDURE

Part A

Thoroughly clean a copper strip and carbon rod with steel wool or sandpaper, and immerse them in a saturated solution of copper sulfate (Figure 41-1). Connect the positive pole of a d-c source of current to the copper electrode and the negative pole to the carbon rod. With an ammeter (0-30 amps) connected in series in the circuit allow the current to flow for five minutes, taking current readings every thirty seconds.

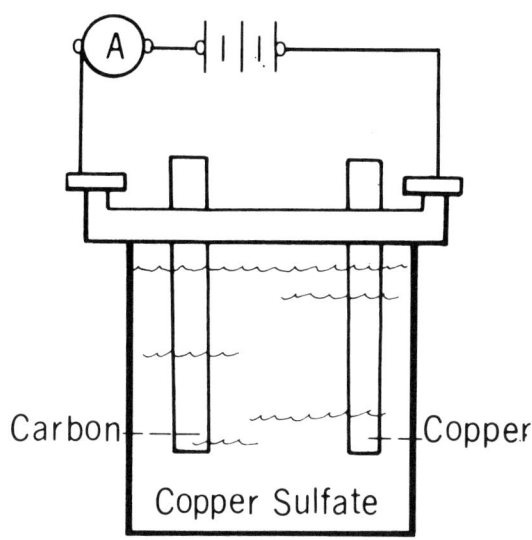

FIGURE 41-1

Time in Minutes	.5	1	1.5	2	2.5	3	3.5	4	4.5	5
1. Current	__	__	__	__	__	__	__	__	__	__

2. What is the average current during the five minutes?
3. Describe the appearance of the carbon rod.
4. How many coulombs of charge were transferred during

the plating process? (You may refer to a textbook if the relationship of ampere-seconds and coulombs is not known.)
5. How many electrons is this?
6. How many cupric ions (+ 2 valence) were plated?
7. What part of a mole (6.024×10^{23} atoms) is this?
8. What weight of copper was plated?

Part B

Reverse the polarity of the cell used in Part A and allow the current to flow for 3-5 minutes.
1. What is happening at the carbon rod?
2. With the same average current as used in Part A what time would be required to completely unplate the copper? Explain.

DISCUSSION OF OUTCOMES

Part A

1. Student answer.
2. Student answer derived by dividing the total amperes of current in item 1 by the number of recorded readings.
3. Copper is plated upon the carbon rod.
4. Student answer. Coulombs of charge = amperes of current $\times$ time in seconds.
5. Student answer. No. of electrons =
    number of coulombs $\times$ 6.28 $\times$ 10^{18} $\dfrac{\text{electrons}}{\text{coulombs}}$
6. Student answer. No. of cupic ions neutralized = $\dfrac{\text{no. of electrons}}{2}$
7. Student answer. Moles of copper = $\dfrac{\text{no. of copper atoms (ions) plated}}{6.024 \times 10^{23}}$

Optional Problems 185

8. Student answer. Weight in grams of copper plated = moles of copper plated × atomic weight of copper.

Part B

1. The copper atoms are being ionized and are returning to the solution.
2. It should take the exact same time to unplate the copper as it took to plate it with the same average current. The number of electrons required for the unplating is equal to the number that was required for the plating.

EXERCISE 42
OPTIONAL PROBLEMS TO TEST LABORATORY SKILLS

(Teacher) Exercise 42 is comprised of two problems which are designed to test the laboratory skills of the students. The teacher should operate as consultant and advisor while allowing the students the greatest degree of independence possible. At this point, if the series of laboratory exercises previously described have been utilized at all sequentially, the capable independent operators in each class will have been identified. It may be well to reorganize the work groups in such a way to place at least one capable investigator in each group. This procedure will not affect the behavior of the unwilling investigator, but the methods and thought patterns of the capable performer may serve as a model to the willing but less capable student.

(Student) It is again time to put to use those skills which you have acquired in the physics laboratory. The culminating skill which each student should try to develop is the ability to integrate all of his acquired knowledge and individual skills in seeking an answer to a problem with which he is faced. The challenge is great, but the rewards in terms of satisfaction are greater.

Below are described three problems from which you may choose for your independent investigation. Your choice of a problem should provide for you and your partners the greatest possible challenge which you are able to meet. If you have doubts about which of the problems is best for your particular group consult with your teacher.

PROCEDURE

Problem A—The Intensity of Radiation

Most radiation which is ejected from a radioactive substance carries an electrical charge. The intensity of radiation in an area may be estimated by the rate at which an electroscope is discharged. Can you determine the relationship between the intensity of radiation from a radioactive material and distance from the source of the radiation?

Problem B—Estimating the Force Between Two Charged Spheres

A charged sphere suspended on a non-conducting silk thread of length L isolated in space will be acted upon by the force mg (its weight). A second charged sphere which is placed in proximity to the suspended sphere will result in a force (f) which acts upon the suspended sphere causing it to be displaced by a distance (d). Can you determine the magnitude of the force which exists between the two charged spheres?

Problem C—The Measurement of Electric Current

The magnitude of a current through an electrical circuit is commonly measured utilizing two of its properties: 1) repulsion and/or attraction of the magnetic field about the current-bearing conductor, or 2) the rate or amount of a metal which is plated in an electrochemical cell. Can you construct and calibrate an instrument which may be used to measure the flow of an electric current?

The suggestions for investigating the three problems are in

Optional Problems

no way binding. Students should feel free to search available literature for alternate suggestions or to devise processes of investigation of their own design.

DISCUSSION OF OUTCOMES

Problem A

The student investigators may employ an electroscope or, if available, a radiation meter to measure the relative intensities of radiations at various dirstances from a radioactive source. The first step in the investigation should be to establish the "normal" discharge rate of the electroscope due to leakage and "background" radiation. Then the discharge rate at various distances may be established. Putting the data into tabular and/or graphical form will aid the student in establishing the inverse square law as the intensity-distance relationship. They may require a reminder that intensity of radiation is inversely related to the time of discharge of the electroscope.

Problem B

A suggested apparatus and force diagram appears in Figure 42-1. As corresponding sides of similar triangles the ratio $\frac{F}{Mg}$ equals the ratio $\frac{d}{1}$. Therefore the force between the spheres (f) is equal to $\frac{Mg\ d}{1}$. The force in dynes is equal to the product of the mass (grams), the acceleration constant $\frac{(980\ cm)}{sec^2}$, and the distance (cm), divided by the length (cm.). To prevent sideward displacement due to the addition of vector forces as the charged spheres are being repelled, it may be necessary to attach silk threads as guidelines on the suspension thread.

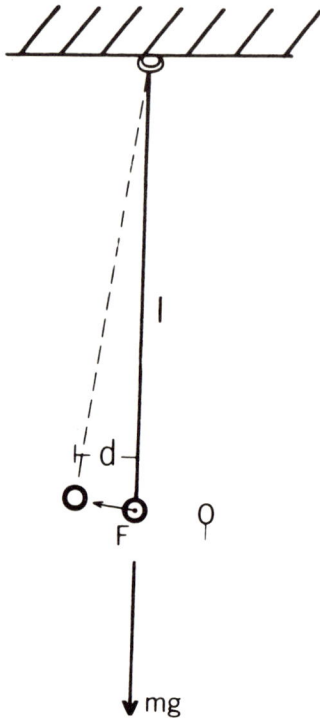

FIGURE 42-1

Problem C

Review of physics textbooks might suggest several possible alternatives for measuring electric current. Careful weighing of the amount of a metal deposited over a period of time is one method. The suspension of two straight conductors in an evaporating dish filled with mercury is a second. A current passing through the conductors (via the mercury) will result in magnetic forces of repulsion. The strength of the repulsion is proportional to the strength of the current. A third suggestion is to suspend a soft iron bolt or nail by a rubber band in the center of a coil of insulated wire (see Figure 42-2).

Optional Problems

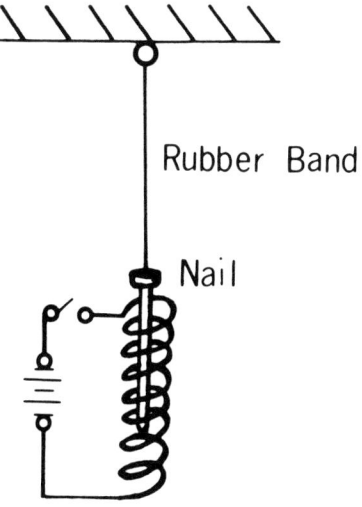

FIGURE 42-2

The magnetic field created by the passage of a current through the wire coil will draw the nail downward into the core of the coil. The magnitude of the current through the coil may be regulated by controlling the voltage which is applied to the current.

Unit VI

Circuits and Magnetism

The Wheatstone Bridge
Series Circuits
Parallel Circuits
Series and Parallel Circuits Combined
Electrical Power
Magnetic Fields
Magnetic Fields About Current Bearing Conductors
Induced Currents
The Transformer
Inductance and Capacitance

Unit VI

EXERCISE 43—THE WHEATSTONE BRIDGE

APPARATUS

1. Wheatstone Bridge
2. Contact Switch
3. Galvanometer
4. Two resistances of unknown value
5. A resistance box
6. A resistor of known value

(Teacher) The Wheatston Bridge has long been a classic exercise in elementary physics laboratories. The design of the apparatus employs some very fundamental concepts of current electricity. The student upon seeing current flow through the galvanometer in either of two directions, or not at all, depending upon the length of an adjustable resistor, is often brought to real understanding of the interrelationships of the factors of current, voltage, and resistance for the first time. The process skills required for this exercise would appear to be understanding number relationships (simple ratios) and experimenting. The student is also asked to draw a deductive conclusion regarding the relationship of current through the galvanometer and voltage differences.

(Student) The inaccuracies that occur due to power loss in the ammeter-voltmeter method of determining resistance may be eliminated in instruments employing the Wheatstone Bridge principle.

PROCEDURE

The schematic diagram of Figure 43-1 represents a Wheatstone Bridge. R_x is a resistor of unknown value. The value of R_3

is known. The ratio of the resistance values $\dfrac{R_1}{R_2}$ is the same as that of the lengths of the resistors $\dfrac{L_1}{R_2}$.

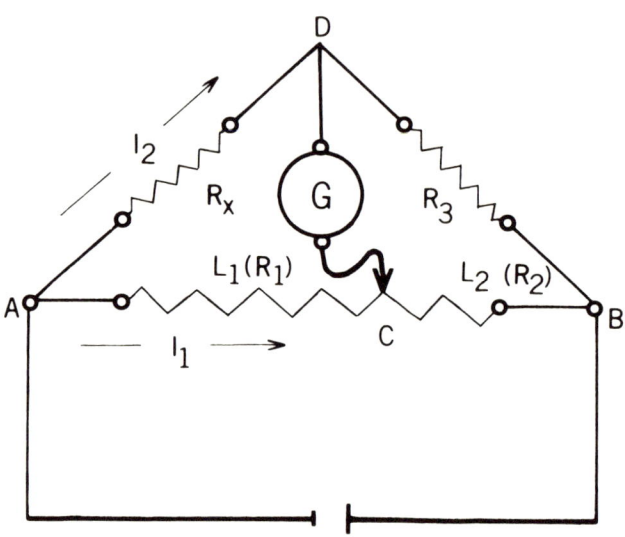

FIGURE 43-1

1. Why is this known to be true?
2. If a current flows through the galvanometer with the switch closed what does this indicate concerning the potential difference between points C and D?
3. If the length of the resistor, ABC, is adjusted so that no current flows through the galvanometer what does this mean concerning the potential difference between points C and D?

If you concluded there is no potential difference between points C and D when no current flows through the galvanometer you are correct. Follow the reasoning from which will evolve the mathematical relationship which permits the use of the Wheatstone Bridge as a quick means of determining the value of an unknown

The Wheatstone Bridge

resistor. When the system is so adjusted that no current is indicated on the galvanometer, the voltage drop across AD must equal the voltage drop across AC and the voltage drop across DB must equal that across CD.

If: $V_{AD} = V_{AC}$ If: $V_{DB} = V_{CB}$
Then: $I_2 R_x = I_1 R_1$ Then: $I_2 R_3 = I_1 R_2$

Dividing one expression by the other we get: $\dfrac{R_x}{R_3} = \dfrac{R_1}{R_2}$

Since R_1 and R_2 are lengths of a uniform resistance wire $\dfrac{R_1}{R_2} = \dfrac{l_1}{l_2}$ therefore the expression may be written: $R_x = \dfrac{l_1}{l_2} R_3$

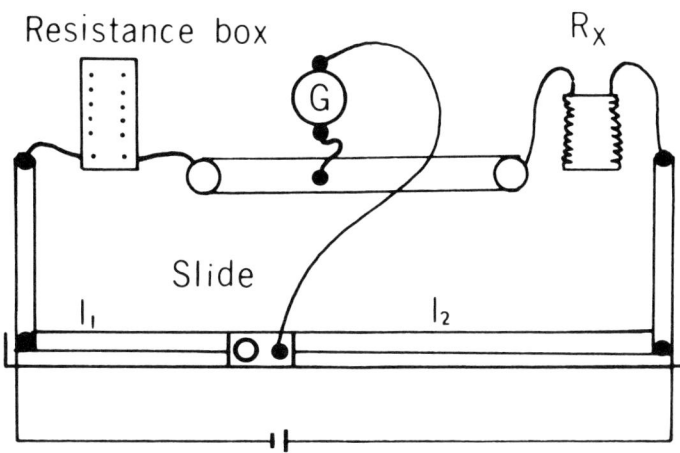

FIGURE 43-2

Set up the Wheatstone Bridge as shown in Figure 43-2 and adjust the slide until the galvanometer registers zero and record the following:

4. $R_3 =$ _____.
5. $l_1 =$ _____.
6. $l_2 =$ _____.
7. The value of R_x is _____.
   Replace R_x with another unknown resistance and find its value.

8. $R_3 =$ _____.
9. $l_1 =$ _____.
10. $l_2 =$ _____.
11. The value of R_x is _____.
12. Explain in your words the principle of Wheatstone's Bridge.

DISCUSSION OF OUTCOMES

1. The resistance of a uniform, straight conductor is proportional to its length.
2. If a current flows through the galvanometer this indicates a potential difference exists between points C and D.
3. No current flowing through the galvanometer indicates no potential difference exists between points C and D.
4. Student data.
5. Student data.
6. Student data.
7. Student answer found by employing $R_x = \dfrac{l_1}{l_2 \; R_3}$
8. Student answer.
9. Student answer.
10. Student answer.
11. Student answer found by employing $R_x = \dfrac{l_1}{l_2 \; R_3}$
12. Student answer.

EXERCISE 44—SERIES CIRCUITS

APPARATUS

1. Three lamp sockets
2. Three lamps of equal wattage
3. Two lamps of different wattages
4. Knife Switch
5. Voltmeter (to exceed emf of source)
6. Ammeter (0-3 amps)

Series Circuits

(Teacher) The understanding of the relationships of electromotive force, current, and resistance in electric circuits is fundamental to any appreciation of the technological application of electrical theory. This exercise dealing with the effects of resistors in series allows the student to experimentally verify or discover basic relationships of current electricity. The relationships are gained as the product of both deductive and inductive manipulations of experimental data. Fundamental processes required of the students would appear to be: understanding numerical relationships, interpreting data, and formulating hypotheses.

(Student) The magnitude of the current which flows through a circuit which is connected to some source of emf is determined by the value of the resistance of the circuit. The manner in which resistors are joined determines the total or equivalent resistance of the circuit. When resistors are connected one after the other so that the current has only one continuous path they are said to be connected in series.

PROCEDURE

Part A

With a bank of three light sockets in series (as in Figure 44-1) which contains bulbs of equal wattage connected to a source of emf, it is possible to investigate the properties of resistances in series.

1. With a voltmeter determine the voltage drop across bulb A by connecting a voltmeter to posts a_1 and a_2.

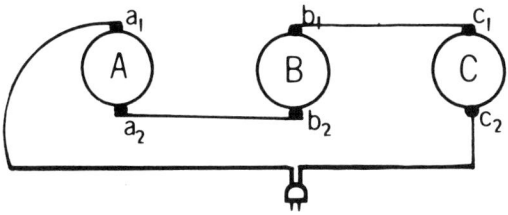

FIGURE 44-1

Connect the voltmeter to b_1, and b_2 and c_1, c_2.
1. Voltage drop across the resistor B?
2. Voltage drop across resistor C?
3. How do these compare?
4. Now find the total voltage drop across resistor A plus B by connecting the voltmeter to a_1 b_1.
5. What do you think the total voltage drop across A, B, and C would be?
   Check this by connecting a voltmeter to a_1 c_2.
6. What is the voltage of the source of emf?
7. Make a statement concerning the sum of voltage drops and the emf of the source in series circuits.

Connect an ammeter in series successively between the switch and a_1, between a_2 and b_2, and between b_1 and c_1 and determine the current of each position.
8. How do they compare?
9. Make a statement concerning the current in a series circuit.

Part B

Replace two of the bulbs with bulbs of different wattages and answer the following questions.
1. Voltage drop across A?
2. Voltage drop across B?
3. Voltage drop across C?
4. Voltmeter reading which indicates the total voltage drop across three resistances?
5. Do the sum of the separate voltage drops equal the total drop across the three resistors?
6. Is the total drop across the three resistors equal to that of the source of emf?

Insert an ammeter in series in the three positions described in Part A and record their readings.
7. How do your conclusions regarding the magnitude of the current in Part B compare with your conclusion dealing with the same factor in Part A?
8. Using Ohm's Law find the resistance of resistors A, B, and C, used in Part B.

Series Circuits

9. The sum of the resistances of this series circuit (R_T) is _____ ohms.
10. May the values of the separate resistances in a series circuit be added together to determine a total value for the resistance of a circuit?

DISCUSSION OF OUTCOMES

Part A

1. Student data.
2. Student data.
3. The voltage drop across each of the resistors (lamps) should be approximately equal.
4. Student data. (This value should approximate the sum of the separate values of the two resistors.)
5. Extrapolating the discovery of item 4 that resistances are additive the student should hypothesize the total resistance of the three resistors to be equal to the sum of the values of the separate resistors.
6. Student data.
7. Student statement. The effect of the statement should be that the sum of the voltage drops is equal to the voltage of the source of emf.
8. Student statement. The magnitude of the three ammeter readings should be approximately equal.
9. Student statement. The effect of the statement should be that the current is the same throughout a series circuit.

Part B

1. Student data.
2. Student data.
3. Student data.
4. Student data.
5. The sum of the voltage drops of the three separate resistors should approximate the voltmeter reading across all three resistors.
6. Yes.

7. The conclusion is substantiated that the current is everywhere equal.
8. Student answers found by employing $R = \dfrac{V}{I}$
9. Student answer found by adding together the values of the three resistances from item 8.
10. Yes. The data substantiates the conclusion that the total resistance of a series circuit may be determined by adding together the values of the separate resistances.

EXERCISE 45—PARALLEL CIRCUITS

APPARATUS
1. Three lamp sockets
2. Three lamps of equal wattage
3. Two lamps of other wattages
4. Knife-blade switch
5. Voltmeter (range to exceed the voltage of the source of emf)
6. Ammeter (0-3 amps)

(Teacher) One of the most difficult concepts for beginning physics students to master is the relationship of resistance and current in parallel circuits. This exercise is designed to clarify this relationship by demonstration. The student will witness the result of adding resistors in parallel one at a time to a circuit. It should become apparent to him that the addition of a resistor might also be considered to be providing an additional path through which current may flow, and thus the equivalent resistance of the circuit is lessened. The process skills developed by the exercise would appear to be understanding numerical relationships, interpreting data, experimenting, and formulating hypotheses.

(Student) Resistances connected in parallel provide multiple paths for current. The current divides and flows through each resistance. The proportion of the total current in each resistor is inversely related to the value of the resistance, that is, the branches offering least resistance will carry the greater current.

Parallel Circuits

PROCEDURE

Part A

Place lamps of equal wattage (resistance) in a lamp bank as in Figure 45-1. After closing the switch unscrew one lamp.

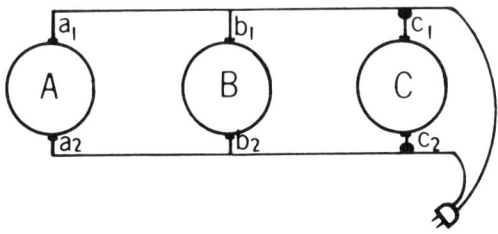

FIGURE 45-1

1. Are the other two lamps affected? Unscrew a second lamp.
2. Is the third lamp affected? After replacing all lamps find the voltage drop across each lamp by placing the leads of a voltmeter successively at c_1c_2, b_1b_2, and $a_1\ a_2$.
3. How do the voltage drops across the lamps compare?
4. How do the voltage drops of each lamp compare with the voltage of the source of emf?

The current through each resistor and the total current of the circuit may be found by placing an ammeter successively in position 1, 2, 3, and 4, as in Figure 45-2.

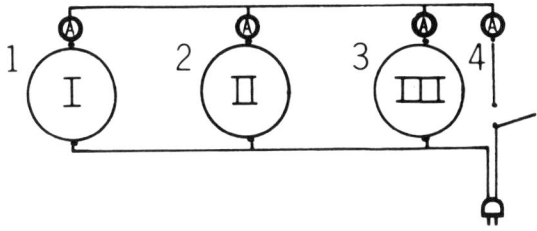

FIGURE 45-2

5. How do the currents through the three lamps compare?
6. What is the relationship of the current through the lamps and the total current?

Part B

Replace two bulbs of the lamp bank with two of different wattages (resistances) and check the voltage drop across each lamp as in Part A.

1. Is the voltage drop across each lamp the same? By placing an ammeter in positions 1, 2 and 3 record the current through the three lamps.
2. Is the current the same through the three lamps?
3. With the ammeter in position 4 record the current in the circuit with:
    a) all bulbs unscrewed,
    b) one burning,
    c) two burning,
    d) with all three burning.
4. How does the addition of each bulb affect the total current in the parallel circuit?
5. If the total current of a parallel circuit increases with the addition of each new resistor this indicates what about the equivalent resistance of the circuit?
6. Use Ohm's Law to find the resistance of each bulb from the preceeding data.
7. Using the relationship $\frac{1}{R_t} = \frac{1}{R_a} + \frac{1}{R_b} + \frac{1}{R_c}$ find the equivalent resistance of the circuit.
8. Using the equivalent resistance and the source voltage compute the total current.
9. How does this compare with the reading of the ammeter in position 4?
10. How would adding an additional resistor in parallel affect the total current?

Series and Parallel Circuits Combined

DISCUSSION OF OUTCOMES

Part A

1. No.
2. No.
3. The voltage drops are the same.
4. The voltage drop across each lamp is equal to the voltage of the source of emf.
5. The current through each lamp is the same.
6. The total current is equal to the sum of the separate currents through the branches.

Part B

1. Yes. The voltage drop across each lamp is the same.
2. No. The current through the lamps is not the same.
3. a) no current, b) student data, c) student data, d) student data.
4. The addition of each bulb increases the total current of the circuit.
5. An increase in current with a constant voltage indicates a reduction in the equivalent resistance of the circuit.
6. Student answer found by employing $R = \dfrac{V}{I}$.
7. Student answer.
8. Student answer found by employing $I = \dfrac{V}{R}$.
9. The results should compare favorably.
10. Adding an additional resistor in parallel would increase the total current of the circuit.

EXERCISE 46
SERIES AND PARALLEL CIRCUITS COMBINED

APPARATUS

1. Three lamps of different wattages and
2. A 120 volt- a-c source of emf, (or)
3. Three resistors of different values and
4. A d-c source of one to six volts

(Teacher) Exercise 46 may be viewed as a culminating activity for Exercises 44 and 45. To be examined in Exercise 46 are the combined effects of series and parallel circuits. The student should come to understand through this exercise that the equivalent resistance of a parallel network may be added to the resistances in series to determine the total resistance of a circuit. Process skills required for the activities of this exercise are: understanding numerical relationships, hypothesizing, and experimenting.

(Student) An electrical circuit that is comprised of resistances in both parallel and series may be studied to ascertain the relationships of resistances, voltages, and currents in the branches of such a circuit.

PROCEDURE

Construct a circuit as shown in Figure 46-1. Connect the ammeter in position A_1. If household current (120 volts a-c) is used three lamps of different wattages should be used and an a-c ammeter (0-30 amps). If a low voltage d-c source is used, use three resistors of known values and a D.C. ammeter (0-1 amps). Position A_1 of the ammeter indicates the total current in the circuit. All current must pass through it.

1. How great is the total current? Now connect the ammeter successively in position A_2 and A_3.

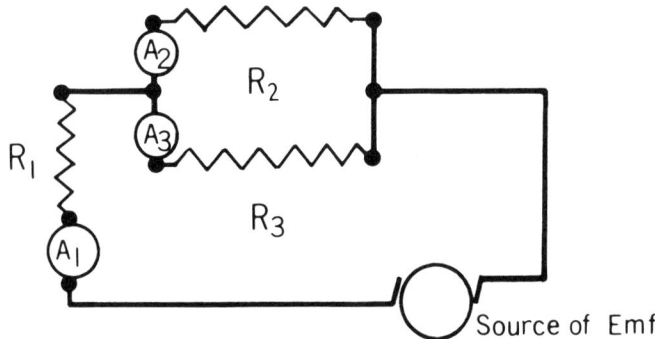

FIGURE 46-1

Series and Parallel Circuits Combined 205

2. What is the current through R_2?
3. What is the current through R_3?
4. How does the sum of the currents in R_2 and R_3 compare with the current in R_1?
5. With a voltmeter check the voltage drop across:
    a) R_1
    b) R_2
    c) R_3
6. What is the voltage of the source of emf?
7. Can you find the value of the resistances?
    a) R_1
    b) R_2
    c) R_3
8. What is the equivalent resistance of the parallel network of the circuit?
9. What is the total resistance in the circuit?
10. According to Ohm's Law the total current should be how great?
11. How does the experimental value for the total current compare with that read on the meter A_1?

DISCUSSION OF OUTCOMES

1. Student data.
2. Student data.
3. Student data.
4. The sum of the currents in R_2 and R_3 should equal the current in R_1.
5. a) Student data, b) student data, c) student data (Note: The voltage drops across R_2 and R_3 should be equal.)
6. Student answer to be derived by measurement.
7. a) student answer, b) student answer, c) student answer. These answers may be computed with the relationship, $R = \dfrac{V}{I}$
8. Student answer to be computed by $\dfrac{I}{R_T} = \dfrac{I}{R_1} + \dfrac{I}{R_2}$

206 *Circuits and Magnetism*

9. Student answer found by adding the equivalent resistance found in item 8 to the value of resistor R_1 found in item 7.
10. Student answer to be found by $I = \dfrac{V}{R}$
11. Student answer. The values should compare favorably.

EXERCISE 47—ELECTRICAL POWER

APPARATUS
1. Two bulbs of different wattage ratings
2. A single receptacle
3. Hot plate (or other appliance)
4. Voltmeter (the range to exceed the emf of the power source)
5. Ammeter (1-3 amps)

(Teacher) The relationships of electric current, energy, and power are important concepts in the students' understanding of technological applications of this energy form. Exercise 47 is designed to provide a base upon which the student may build as he seeks to establish these interrelationships in his own mind. Laboratory skills which are thought to be utilized in performing the exercise are experimenting, understanding numerical relationships and formulating hypotheses.

(Student) The basic unit of electrical power is the watt. One watt is equivalent to the work done by one coulomb of charge moving between a potential difference of one volt in one second of time. Power in watts may be found by multiplying the voltage (volts) times the current (amps). The kilowatt is one thousand watts, and the kilowatt-hour is the amount of electrical energy provided when one thousand watts is consumed continuously for one hour.

PROCEDURE

Part A

Place a single bulb of known wattage in a socket and connect it to a source of household current.

Electrical Power

1. Check the current through the bulb and the voltage across it.
2. What wattage rating would you assign to this bulb?
3. How does this compare to the commercial rating?
4. Explain any difference.

Now insert a bulb with a different wattage rating and repeat the above procedure.

5. a) Current?
   b) Voltage?
   c) Wattage?
6. How does the experimental value for the wattage compare with the commercial rating?

Part B

Connect a hot plate element to a source of emf with an ammeter in series in one line and a voltmeter in parallel across the lines. (The range of the voltmeter should exceed the emf of the power source.)

1. Check the voltage and current at three minute intervals for 15 minutes.

	Current	Voltage
Start	_____	_____
After 3 min.	_____	_____
After 6 min.	_____	_____
After 9 min.	_____	_____
After 12 min.	_____	_____
After 15 min.	_____	_____
Average	_____	_____

2. What is the computed wattage of the hot plate?
3. How many watt-hours of energy were consumed?
4. Killowatt-hours?
5. What is the cost of operating this appliance during this exercise if power costs $.05 per killowatt-hour?
6. How could you find the power rating in watts of an appliance when the current and resistance are known?
7. How could you find the power rating in watts of an appliance when the voltage and resistance are known?

DISCUSSION OF OUTCOMES

Part A

1. Student data.
2. Student answer determined by: $P = VI$
3. Student answer.
4. Differences may be due to inaccuracies of meters and changing characteristics of resistor wires.
5. Student data.
6. Student answer.

Part B

1. Student data.
2. Student answer found by: $P = VI$
3. Student answer found by: watt-hrs. $= P \times 15 \text{ min}/60 \text{ min/hr.}$
4. Student answer found by: $\text{kw-hrs.} = \dfrac{\text{watt-hrs.}}{1000 \text{ watt-hrs/kwhr.}}$
5. Student answer found by: $\text{Cost} = \text{kw-hr} \times \$.05$.
6. $P = I^2 R$
7. $P = \dfrac{V^2}{R}$

EXERCISE 48—MAGNETIC FIELDS

APPARATUS

1. Two bar magnets
2. One piece of stiff paper or cardboard
3. Iron filings

(Teacher) Exercise 48 is an activity quite common to general science classes wherein the general shape of magnetic fields about bar magnets are described by the orientation of iron filings.

Magnetic Fields

Understanding of magnetic fields derived from this sort of activity is extended by the nature of the questions asked of the students to include an appreciation of a construct or scientific model. Hopefully the students will realize that magnetic flux lines, or lines of force, do not actually exist in nature, but are convenient schemes proposed to describe the effects of magnetic fields. The highest level process skills required of the students are making hypotheses and building models.

(Student) Magnetic fields are regions where magnetic effects are observable. We use magnetic flux lines, or lines of force, as devices for describing the orientation and strength of a magnetic field. If a tiny north pole (N) were placed at position X in Figure 48-1, the vector sum of repulsive force L and attractive force A would be R, which defines the orientation and magnitude of the magnetic field at that point. A succession of such points would trace a path from the magnetic north pole to the magnetic south pole that would be known as a line of force or flux line. Obviously magnetic flux lines do not exist as natural phenomena but are man-made contrivances designed to explain and predict the effects of magnetic force fields. Such a procedure is an example of model-building. All observed effects of magnetic fields may thus be weighed against those that would be predicted by the model. In this way the model is tested. A good model provides the investigator with a system for predicting the consequences of a physical condition and a means for testing his observations.

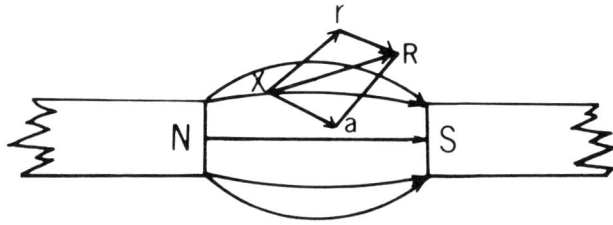

FIGURE 48-1

PROCEDURE

Part A

Place a piece of cardboard or stiff paper cover a bar magnet. Sprinkle iron filings upon the cardboard and tap gently with the forefinger.
1. Make a drawing of the lines of force.
2. In what region of the magnetic field does the magnetic effect appear to be the greatest?

Part B

Place the N pole of one bar magnet 2-3 centimeters away from the S pole of another, cover with a piece of cardboard, and sprinkle on the iron filings.
1. Make a drawing of the lines of force.
2. Do these appear to be continuous flux lines between the N and S poles?

Part C

Place the N pole of one bar magnet 2-3 centimeters away from the N pole of another, cover with the cardboard, and sprinkle on the iron filings.
1. Make a drawnig of the lines of force.
2. Do they appear to be continuous flex lines between the N and N poles?
3. What path do you think an isolated N pole would follow if placed in this magnetic field?

Part D

Place N and S poles on a table with a small soft iron bar between them with a spacing of about two centimeters between the objects. Place a piece of cardboard over them and sprinkle with iron filings.
1. Draw the diagram of the lines of force.
2. What appears to be the effect of placing a soft iron object in a magnetic field?
3. Can you guess why this is so?

About Current Bearing Conductors 211

DISCUSSION OF OUTCOMES

Part A

1. Student diagram.
2. The magnetic field is strongest at the ends of the bar magnet.

Part B

1. Student diagram.
2. There do appear to be continuous flux lines between the poles.

Part C

1. Student diagram.
2. There are no continuous flux lines between the N poles.
3. An isolated N pole would be repelled along a path which approximates the general pattern of the flux lines. (Note: An N pole placed exactly in the center between two N poles of equal strength would remain stationary with the sum of the forces acting upon it being equal to zero.)

Part D

1. Student diagram.
2. The flux lines appear to reside in the soft iron bar.
3. The iron bar becomes a magnet and acts as a continuation of the other magnets. Flux lines appear to remain internal within a magnetic material.

EXERCISE 49
MAGNETIC FIELDS ABOUT CURRENT BEARING CONDUCTORS

APPARATUS

1. Low voltage D.C. source (3 dry cells)
2. One meter of 18 or 20 gauge insulated wire
3. Compass
4. Tacks or paper clips

212   Circuits and Magnetism

(Teacher) One of the most dramatic methods for establishing the intimate association between current electricity and magneic fields is to allow the students to discover some of the relationships himself. Exercise 49 is designed to permit the investigator to establish the rough effect upon the strength of magnetic fields about a current bearing conductor of forming a helix; of adding a highly permeable core, of adding turns to a helix; and of increasing the current through the conductor. The process skills involved are primarily those of experimenting and interpreting data.

(Student) There is a magnetic field formed about any current bearing conductor. The effects of this magnetic field may be concentrated by coiling the wire to form a solenoid, and further concentrated by introducing an iron core to the solenoid.

PROCEDURE

Part A

Lay a small compass upon a table. Across the face of the compass lay part of the length of one meter of 18 or 20 gauge wire which has been connected in series with a contact key and low voltage D.C. source (1-3 volts). Close the contact key for an instant and observe the compass needle.

1. How does it align itself in relation to the wire? Now hold the compass over the wire and close the key for an instant.
2. How does it align itself in relation to the wire?
3. What does this indicate concerning the magnetic field about a straight conductor?

Part B

Coil about 50 turns of the wire about a soft iron core (a large nail will do) and remove the core. Hold the end of the coil close to the compass and close the key for an instant.

1. Is the compass affected?
2. Can you make some statement as to the relative strengths of the magnetic field about the straight and coiled conductors?

    Try to attract a tack or paper clip with the coil.
3. Successful?

Part C

Insert the soft iron core into the 50 turn coil and close the switch.

1. Count the tacks or paper clips the magnet will now attract.
2. The addition of the soft iron core had what effect upon the ability of the magnetic field to attract magnetic material?

Now wrap about 50 more turns about the soft iron core and close the contact key.

3. Count the tacks or paper clips the magnet will attract.
4. What is the effect of a greater number of turns on the strength of an electromagnet?

Part D

Connect the 100 turn coil to a higher D.C. voltage (3-4 volts) and close the contact key.

1. Count the number of tacks or paper clips the electromagnet will attract.
2. What is the effect of increased voltage (current) upon the strength of an electromagnet?
3. Write a statement describing those factors which appear to affect the strength of a magnetic field which lies about a current bearing conductor.

DISCUSSION OF OUTCOMES

Part A

1. The compass aligns at right angles to the conductor.
2. The compass swings 180 degrees and aligns at right angles to the conductor.
3. The magnetic field about a straight, current-bearing conductor is circular in shape and at right angles to the conductor.

Part B

1. Yes.
2. The magnetic field about the coiled conductor is stronger.

3. Usually there is no noticeable effect upon a tack or paper clip.

Part C

1. Student data.
2. The addition of the core greatly enhanced the ability of the magnetic field to attract magnetic materials.
3. Student data.
4. An increase in the number of turns increased the attractive power of the magnetic field.

Part D

1. Student data.
2. An increase in the current through a conductor increases the strength of the magnetic field.
3. The strength of the magnetic field about a current-bearing conductor is increased by: a) forming helixes, b) adding a permeable core, c) increasing the number of turns, and d) increasing the current through the conductor.

EXERCISE 50—INDUCED CURRENTS

APPARATUS

1. Galvanometer
2. Coil of approx. 25 turns of copper wire of a dia. equal to, or exceeding the dia. of 20 gauge wire
3. A coil of 50 turns of similar wire
4. Bar magnet
5. Horseshoe magnet

(Teacher) The concept of electromagnetic induction is fundamental to the structure of electrical theory. In Exercise 50 the student observes the concept in its most primitive form, and tests several variables for their affects upon the strength of an induced current. Finally the student is asked to hypothesize concerning the effect of maintaining the magnet in a fixed position and causing

Induced Currents 215

the solenoid to move through the region of the magnetic field. The process skills involved in the exercise are thought to be those of hypothesizing, experimenting and controlling variables.

(Student) A free electron in the presence of a fluctuating magnetic field will have forces exerted upon it. The work that is done on an electron in moving it from one point to another is the measure of the potential difference, or voltage, between the two points. When the loosely held electrons in a conductor are caused to migrate along the conductor by a changing magnetic field an electric current is the result. A current in a conductor which has resulted from a fluctuating magnetic field is said to be an induced current.

PROCEDURE

Connect the ends of a coil of at least 25 turns of wire to a galvanometer. Quickly insert a bar magnet into the center of the coil and note the movement of the galvanometer needle. Withdraw the bar magnet quickly from the coil.
1. How does the galvonometer react?
2. Can you draw any conclusions as to whether the direction with which a coil intercepts a magnetic field affects the direction of the current?

To investigate factors affecting the strength of an induced current first connect a galvanometer to a coil containing at least twice the number of turns of the previously used coil. Now move the bar magnet through the coil.
3. Is there more current?
4. What was the only variable that was changed?

Through the same coil move the bar magnet first rapidly, then slowly.
5. Which movement produces the greater induced current?
6. Write a statement concerning the effect of the rate of movement of a magnetic field upon the induced currrent.

Induce a current in the coil with a strong horseshoe magnet.
7. Write a statement concerning the effect of the strength of a magnetic field upon the current induced by the magnetic field.

8. Does it make a difference as to its ability to induce a current if the magnet is stationary and the coil moves? Check your hypothesis experimentally.

DISCUSSION OF OUTCOMES

1. The galvanometer moves in a given direction upon the insertion of the bar magnet into the center of the coil and the opposite direction upon its withdrawal.
2. The direction of the current flow is determined by the direction of the movement of the bar magnet.
3. There is a greater current.
4. The only variable changed was the number of turns of the coil.
5. The rapid movement produces the greatest current.
6. The more rapidly a magnetic field passes over a conductor coil, the greater is the induced current.
7. The stronger a magnetic field the greater will be the current it induces in a conductor coil, all other variables remaining constant.
8. The current is equally great when the magnet remains stationary and the conductor coil moves through the magnetic field.

EXERCISE 51—THE TRANSFORMER

APPARATUS

1. A cardboard tube with a coil of 25 turns at one end and a coil of 250 turns at the other
2. Dry cell or other d-c source
3. Galvanometer
4. Iron Wire
5. Flashlight bulb and socket
6. A-C ammeter and/or voltmeter

(Teacher) The electrical transformer is not only a commercially useful device; it may also be used to demonstrate principles

The Transformer

of electromagnetic induction for beginning physics students. In Exercise 51, the students will examine the characteristics of a primitive transformer consisting of two coils of wire wound around a cardboard tube with a bundle of iron wire as a core. Unfortunately the low efficiency of such a device generally precludes the possibility of the students establishing by discovery the relationship of transformer turns and voltage. As a concluding activity the students are asked to experiment with varying numbers of wire turns in the secondary coil of the transformer. In this way they may reinforce their knowledge of the relationships of induction, voltages, and current. The process skills required for completing the exercise would appear to be those of experimenting and interpreting data.

(Student) When a magnetic field is caused to expand or collapse about coils of conducting wire a current is induced to flow in the conductor. A *transformer* consists of two coils, one called the *primary* through which an alternating current flows, and one termed the *secondary* which has a voltage induced upon it by the fluctuating magnetic field of the primary. A similar device with an intermittant direct current in the primary is called an induction coil. The efficiency of the transformer is increased by having both coils wound around a common core of high permeability. The purpose of a transformer is to raise or lower the voltage of a source of alternating current. The ratio of the voltages across the two coils of a transformer is the same as the ratio of the number of turns of each.

$$\frac{\text{Primary voltage}}{\text{Secondary voltage}} = \frac{\text{Primary turns}}{\text{Secondary turns}}$$

PROCEDURE

Part A

Connect a 1-2 volt d.c. source in series with a contact switch and the 25 turn coil of the homemade transformer, as shown in Figure 51-1. To the 250 turn coil attach a galvanometer. Close the contact key, hold it down for an *instant* then release it.

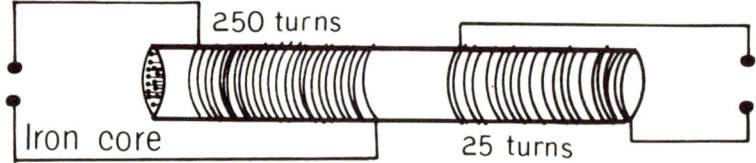

FIGURE 51-1

1. Does the galvanometer indicate an induced voltage at the instant the key is closed?
2. When it remains closed?
3. When it is opened?
4. Does the current flow in the same or opposite directions at the instances when the switch is opened and closed?
5. Explain this series of observations.

Part B

A step-down transformer may be demonstrated by connecting a 120 volt a-c source in series with a knife blade switch and the 250 turn coil. To the 25 turn coil attach a flashlight bulb or a multirange a-c ammeter or voltmeter. (Note: Be sure to start with the most gross range of the meter and work down.) Close the switch an instant and note the reading on the meter.

1. Is a voltage or current detected in the secondary coils? Remove the iron wire core.
2. Is a current or voltage detectable?
3. Conclusions?

Try adding varying numbers of turns of conductor wire to the secondary coil.

4. How does this variable affect the voltage output of the secondary?

Part C

It has been established that induction occurs in the secondary coil of a transformer as a product of the changing current in the primary with its fluctuating magnetic field.

The Transformer 219

1. Considering one cycle of a-c current in the primary coil during which the current increases to a maximum in one direction, decreases to zero, increases to a maximum in the opposite direction and decreases to zero, at what points would one expect to find the greatest amount of induction taking place in the secondary coil?
2. The least induction?

DISCUSSION OF OUTCOMES

Part A

1. Yes.
2. No.
3. Yes.
4. The current flows in opposite directions.
5. Induction is a product of a changing magnetic field. When the switch is closed the current and its magnetic field build to a maximum value, and remain at that value until the switch is opened. When the switch is opened the current and its magnetic field decay to a value of zero. At the closing of the switch the magnetic field of the primary collapses across the secondary coils inducing electrons to flow in the opposite direction.

Part B

1. A voltage or current should be detectable in the secondary.
2. No. Without the iron core no voltage or current is generally detectable in the secondary coil.
3. The student should conclude an iron wire core increases the efficiency of a transformer.
4. The voltage output of the secondary coil is directly proportional to the number of turns of wire.

Part C

1. The greatest amount of the induction will take place at

the instant of the most rapid change in the current. This occurs when the current is zero and is changing directions.
2. The least amount of induction (zero) will occur when the current in the primary coil reaches its maximum value.

EXERCISE 52—INDUCTANCE AND CAPACITANCE

APPARATUS
1. Power source capable of providing 60-120 volts of emf, both a-c and d-c
2. 60 watt bulb and socket
3. choke coil
4. 2 capacitors of about 1 mf
5. Ammeter (0-1 amps)
6. Double throw switch

(Teacher) Exercise 52 provides visual evidence of the effects of induction coils and capicitors in electric circuits. Thus, it serves the very useful purpose of making concrete two difficult-to-visualize abstractions. Inductive and capacitive reactances are important factors in the engineering of electrical distribution systems and electronics equipment. Designers must consider the net effect of resistance, inductance, and capacitance when attempting to obtain maximum transfer of energy between two electric circuits. Another application of these two phenomena is found in the system used to filter electronic signals on the basis of differences in their frequencies. The process skills which are required of the student in this exercise would appear to be experimenting and interpreting data.

(Student) A d-c current through a coil is opposed only by the resistance of the coil wire, but an a-c current through a coil is opposed both by resistance of the coil wire and the reactance which results from the self-induction of the coil. Since *inductive reactance* is equal to $2\pi fL$, the greater the frequency of the alternating current the greater the reactance (f = frequency, L = inductance of the coil).

Inductance and Capacitance

A d-c current will not flow through a capacitor; the reactance is said to be infinite. An a-c current will appear to flow in a circuit containing a capacitor by alternately charging and discharging the plates. Since the reactance of any capacitor in a circuit is determined by the expression $\frac{1}{2\pi fc}$ it follows that the greater the frequency of the a-c current the less is the capacitive reactance (f = frequency, c = capacitance of the capacitor).

PROCEDURE

Part A

Connect a 60 watt bulb in series with a choke coil and a double throw switch so that the circuit may be energized by either a-c or d-c current. (See Figure 52-1.) Throw the switch to the d-c source and note the brightness of the lamp. Now remove the core of the choke slowly.

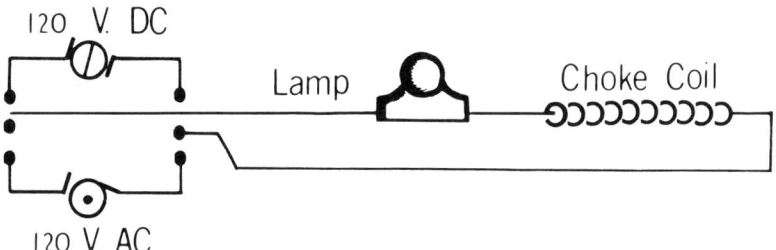

FIGURE 52-1

1. Is the brightness of the lamp affected?
2. Explain.

Now replace the core and throw the switch to the a-c source.

3. Does the lamp glow as brightly as with d-c? Slowly remove the core.
4. Is the brightness of the bulb affected?
5. Explain.

Part B

Replace the choke coil with a one microfarad condenser and throw and switch to d-c current.

1. Result?

Insert an ammeter (0-1 amps) in series with the condenser and throw the switch to the a-c source.

2. Does the bulb light up?
3. What is the reading on the ammeter?

Now join two capacitors in parallel and connect the pair in series in the circuit.

4. With the switch at the a-c position what is the ammeter reading?
5. Explain whether the capacitance was increased or decreased by adding the capacitors in parallel.

Insert the two capacitors in series with each other and with the rest of the circuit.

6. How does this affect the current through the circuit?

Part C

Connect the one microfarad condenser and the choke coil in series with the 60 watt bulb and ammeter.

1. With the switch at the a-c position what is the ammeter reading?
2. Is this greater or less than for the capacitor alone? (Refer to part b)
3. Explain.

DISCUSSION OF OUTCOMES

Part A

1. No, the brightness of the lamp is not affected.
2. There is no inductance in the coil with a d-c power source.
3. No, the lamp does not burn as brightly.
4. The bulb becomes brighter as the core is removed.

5. As the core is removed the inductance (inductive reactance) of the coil is reduced.

Part B

1. The bulb does not light.
2. Yes, the bulb gives off light.
3. Student data.
4. Student data. The ammeter reading should be higher than in item 3.
5. Since $X_c = \dfrac{1}{2\pi fc}$ an increase in current indicates a reduction in X_c which resulted from an increase in capacitance.
6. Joining the capacitances in series decreased the total capacitance, causing an increase in X_c and a corresponding decrease in the current.

Part C

1. Student data.
2. The ammeter reading is less with the choke coil in the circuit.
3. The inductive reactance of the choke coil adds to the total opposition to the flow of current (impedance) of the circuit.